CMOS IMAGERS

CMOS Imagers

From Phototransduction to Image Processing

Edited by

Orly Yadid-Pecht

Ben-Gurion University,
Beer-Sheva, Israel

and

Ralph Etienne-Cummings

Johns Hopkins University,
Baltimore, U.S.A.

KLUWER ACADEMIC PUBLISHERS
BOSTON / DORDRECHT / LONDON

A C.I.P. Catalogue record for this book is available from the Library of Congress.

ISBN 1-4020-7961-3 (HB)
ISBN 1-4020-7962-1 (e-book)

Published by Kluwer Academic Publishers,
P.O. Box 17, 3300 AA Dordrecht, The Netherlands.

Sold and distributed in North, Central and South America
by Kluwer Academic Publishers,
101 Philip Drive, Norwell, MA 02061, U.S.A.

In all other countries, sold and distributed
by Kluwer Academic Publishers,
P.O. Box 322, 3300 AH Dordrecht, The Netherlands.

Printed on acid-free paper

Cover design by Mor Pecht

Printed in the Netherlands.

Dedication

To our loved ones.

Contents

Contributing Authors

Pamela A. Abshire
University of Maryland, College Park, MD, USA

Gert Cauwenberghs
Johns Hopkins University, Baltimore, MD, USA

Matthew A. Clapp
Johns Hopkins University, Baltimore, MD, USA

Marc Cohen
University of Maryland College, Park, MD, USA

Ralph Etienne-Cummings
Johns Hopkins University, Baltimore, MD, USA

Alexander Fish
Ben-Gurion University, Beer-Sheva, Israel

Viktor Gruev
Johns Hopkins University, Baltimore, MD, USA

Honghao Ji
University of Maryland, College Park, MD, USA

Igor Shcherback
Ben-Gurion University, Beer-Sheva, Israel

Orly Yadid-Pecht
Ben-Gurion University, Beer-Sheva, Israel

Preface

The idea of writing a book on CMOS imaging has been brewing for several years. It was placed on a fast track after we agreed to organize a tutorial on CMOS sensors for the 2004 IEEE International Symposium on Circuits and Systems (ISCAS 2004). This tutorial defined the structure of the book, but as first time authors/editors, we had a lot to learn about the logistics of putting together information from multiple sources. Needless to say, it was a long road between the tutorial and the book, and it took more than a few months to complete. We hope that you will find our journey worthwhile and the collated information useful.

The laboratories of the authors are located at many universities distributed around the world. Their unifying theme, however, is the advancement of knowledge for the development of systems for CMOS imaging and image processing. We hope that this book will highlight the ideas that have been pioneered by the authors, while providing a roadmap for new practitioners in this field to exploit exciting opportunities to integrate imaging and "smartness" on a single VLSI chip. The potential of these smart imaging systems is still unfulfilled. Hence, there is still plenty of research and development to be done.

We wish to thank our co-authors, students, administrative assistants, and laboratory co-workers for their excitement and enthusiasm for being involved in this project. Specifically, we would like to thank Alex Belenky, Rachel Mahluf-Zilberberg, and Ruslan Sergienko from the VLSI Systems Center at Ben-Gurion University.

We also would like to thank our mentors, Eric Fossum, Jan van der Spiegel, Albert Theuwissen, Mohammed Ismail, Dan McGrath, Eby

Friedman, Andreas Andreou, Norman Kopeika, Zamik Rosenwaks, Irvin Heard, and Paul Mueller for their support at different stages of this project.

Furthermore, we would like to thank our copy-editor, Stan Backs of SynchroComm Inc.

In addition, we would like to thank our publishers, Kluwer Academic Publishers, and especially Mark de Jongh for being patient with us all the way.

Last but not least, we would like to thank our loved ones for their support during the process. We hope the missing hours with them are worth the result.

Orly Yadid-Pecht and Ralph Etienne-Cummings

Introduction

This book starts with a detailed presentation of the basic concepts of photo transduction, modeling, evaluation, and optimization of Active Pixel Sensors (APS). It continues with the description of APS design issues using a bottom-up strategy, starting from pixels and finishing with image processing systems. Various focal-plane image processing alternatives either to improve imaging or to extract visual information are presented. The book closes with a discussion of a completely non-traditional method for image noise suppression that utilizes floating-gate learning techniques. The final three chapters in fact provide a glimpse into a potential future of CMOS imaging and image processing, where concepts gleaned from other disciplines, such biological vision, are combined with alternative mixed-signal computation circuits to perform complex visual information processing and feature extraction *at the focal plane*. This benefit of CMOS imaging and image processing is still largely unexploited by the commercial sector.

The first chapter reviews the background knowledge and concepts of silicon-based photo transduction, and introduces relevant concepts from semiconductor physics. Several silicon-based photo detectors are examined, including the photodiode and the photogate. This chapter also describes the operation of the charge-coupled device (CCD) imager, the predominant technology available for digital imaging. CCD technology is compared with a promising alternate technology, the APS imager. In addition, the functional performances of several basic pixel structures are compared by considering them as communication channels and determining their ability to convey information about an incident optical signal. At 30 frames per second, information rates are similar for charge-, voltage-, and current-mode pixels.

Comparable trends are found for their information capacities as the photocurrent varies.

The second chapter deals with the modulation transfer function (MTF) of an APS. MTF is one of the most significant factors determining the image quality. Unfortunately, characterization of the MTF of semiconductor-based focal-plane arrays (FPA) has been typically one of the more difficult and error-prone performance testing procedures. Based on a thorough analysis of experimental data, a unified model has been developed for estimation of the MTF of a general CMOS active pixel sensor for scalable CMOS technologies. The model covers the physical diffusion effect together with the influence of the geometrical shape of the pixel active area. Excellent agreement is reported between the results predicted by the model and the MTF calculated from the point spread function (PSF) measurements of an actual pixel. This fit confirms the hypothesis that the active area shape and the photocarrier diffusion effect are the determining factors of the overall MTF behavior of CMOS active pixel sensors, thus allowing the extraction of the minority-carrier diffusion length.

The third chapter deals with photoresponse analysis and pixel shape optimization for CMOS APS. A semi-analytical model is developed for the estimation of the photoresponse of a photodiode-based CMOS APS. This model is based on a thorough analysis of experimental data, and incorporates the effects of substrate diffusion as well as geometrical shape and size of the photodiode active area. It describes the dependence of pixel response on integration photocarriers and on conversion gain. The model also demonstrates that the tradeoff between these two conflicting factors can lead to an optimal geometry, enabling the extraction of a maximal photoresponse. The dependence of the parameters on process and design data is discussed, and the degree of accuracy for the photoresponse modeling is assessed.

The fourth chapter reviews APS design from the basics to more advanced system-on-chip examples. Since APS are fabricated in a commonly used CMOS process, image sensors with integrated "intelligence" can be designed. These sensors are very useful in many scientific, commercial and consumer applications. Current state-of-the-art CMOS imagers allow integration of all functions required for timing, exposure control, color processing, image enhancement, image compression, and analog-to-digital conversion (ADC) on the same die. In addition, CMOS imagers offer significant advantages and rival traditional charge-coupled devices in terms of low power, low voltage and monolithic integration. The chapter presents different types of CMOS pixels and introduces the system-on-chip approach, showing examples of two "smart" APS imagers: a smart vision system-on-chip and a smart tracking sensor. The former is based on a photodiode APS with linear output over a wide dynamic range, made possible by random access to each pixel in the array and by the insertion of additional circuitry into the pixels. The latter is a smart tracking sensor employing analog non-linear winner-take-all (WTA) selection.

The fifth chapter discusses three systems for imaging and visual information processing at the focal plane, using three different representations of the incident photon flux density: current-mode, voltage-mode, and mixed-mode image processing. This chapter outlines how spatiotemporal image processing can be implemented in current and voltage modes. A computation-on-readout (COR) scheme is highlighted. This scheme maximizes pixel density and multiple processed images to be produced in parallel. COR requires little additional area and access time compared to a simple imager, and the ratio of imager to processor area increases drastically with scaling to CMOS technologies with smaller feature size. In some cases, it is necessary to perform computations in a pixel-parallel manner while still retaining the imaging density and low-noise properties of an APS imager. Hence, an imager that utilizes both current-mode and voltage-mode imaging and processing is presented. However, this mixed-mode approach has some limitations, and these are described in detail. Three case studies show the relative merits of the different approaches for focal-plane analog image processing.

The last chapter investigates stochastic adaptive algorithms for on-line correction of spatial non-uniformity in random-access addressable imaging systems. An adaptive architecture is implemented in analog VLSI, and is integrated with the photo sensors on the focal plane. Random sequences of address locations selected with controlled statistics are used to adaptively equalize the intensity distribution at variable spatial scales. Through a logarithmic transformation of system variables, adaptive gain correction is achieved based on offset correction in the logarithmic domain. This idea is particularly attractive for compact implementation using translinear floating-gate MOS circuits. Furthermore, the same architecture and random addressing provide for oversampled binary encoding of the image resulting in an equalized intensity histogram. The techniques can be applied to a variety of solid-state imagers, such as artificial retinas, active pixel sensors, and infrared sensor arrays. Experimental results confirm gain correction and histogram equalization in a 64×64 pixel adaptive array.

We hope this book will be interesting and useful for established designers, who may benefit from the embedded case studies. In addition, the book might help newcomers to appreciate both the general concepts and the design details of smart CMOS imaging arrays. Our focus is on the practical issues encountered in designing these systems, which will always be useful for both experienced and novice designers.

Chapter 1

FUNDAMENTALS OF SILICON-BASED PHOTOTRANSDUCTION

Honghao Ji and Pamela A. Abshire
Department of Electrical and Computer Engineering / Institute for Systems Research
University of Maryland
College Park, MD 20742, USA

Abstract: This chapter reviews background knowledge and concepts of silicon-based phototransduction. Relevant concepts from semiconductor physics, imaging technology, and information theory are introduced. Several silicon-based photodetectors are examined, including the photodiode and the photogate. This chapter also describes the operation of the charge-coupled device (CCD) imager, the predominant technology available for digital imaging. CCD technology is compared with a promising alternate technology, the active pixel sensor (APS) imager. In addition, several basic pixel structures are compared in terms of their functional performance by considering them as communication channels and determining their ability to convey information about an incident optical signal. At 30 frames per second, information rates are similar for charge-, voltage-, and current-mode pixels as the photocurrent varies. Comparable trends are found for their information capacities as the photocurrent varies under idealized operating conditions.

Key words: Photodetector, photoconductor, photodiode, phototransistor, photogate, quantum efficiency, noise, charge-coupled device (CCD), CMOS image sensor, active pixel sensor (APS), information rate, capacity.

1.1   Introduction

Modern photography is a versatile and commercially important technology with numerous applications, including cinematography, spectrography, astronomy, radiography, and photogrammetry. This chemical technology transduces light into a physical representation through a sequence of chemical reactions of silver halide films, including exposure to light, development using benzene derivatives, and fixation using sodium thiosulphate. Generating copies of the original image requires essentially the same procedure. In contrast with conventional film photography, electronic imaging represents light electronically by directly transducing optical signals into electronic signals using image sensors. Such electronic representations enable many applications because processing, storing, and transmitting electronic images are all much easier and more readily automated than comparable manipulations of the physical material representation generated by conventional film photography.

This chapter reviews fundamental aspects of phototransduction using semiconductor image sensors. Concepts of solid-state physics useful for understanding the operation of photodetectors are discussed, and several common silicon-based photodetectors are described. The two predominant image sensor technologies are introduced: charge-coupled device (CCD) technology and active pixel sensor (APS) technology. These are compared and contrasted with each other to understand the advantages and opportunities of each technology. Three different CMOS pixel structures are then compared as communication channels by determining the ability of each one to convey information about optical signals.

1.2   Background physics of light sensing

While a detailed exposition of the interaction between light and matter is beyond the scope of this chapter, a solid understanding of semiconductor physics and the interaction between semiconductors and light will provide insight into the operation of semiconductor imagers. For further details, excellent references on this subject are available [1–4].

1.2.1   Energy band structure

Understanding how optical information is transduced into electronic information requires understanding the electronic properties of semiconductors, starting with the behavior of electronic carriers in semiconductors. From quantum mechanics, it is known that electrons bound to the nucleus of an isolated atom can only have discrete energy levels separated by forbidden gaps where no energy level is allowed [5]. Only one electron can occupy each quantum state, so identical energy levels for two isolated atoms are split into two similar but distinct energy levels as those isolated atoms are brought together. When many atoms come together to form a crystal, the previously discrete energy levels of the isolated atoms are spread into continuous bands of energy levels that are separated by gaps where no energy level is allowed. The energy band structure of electrons in a crystalline solid describes the relationship between the energy and momentum of allowed states, and is determined by solving the Schrödinger equation:

$$\left[-\frac{\hbar}{2m}\nabla^2 + V\left(\vec{r}\right) \right] \psi_k\left(\vec{r}\right) = E_k \psi_k\left(\vec{r}\right)$$
(1.1)

where $\hbar$ is the reduced Planck constant ($\hbar = h/2\pi$), m is the mass of particle, $V\left(\vec{r}\right)$ is the potential energy for an electron, E_k is the total energy, and $\psi_k\left(\vec{r}\right)$ is the wave function.

According to the Bloch theorem, if the potential energy $V\left(\vec{r}\right)$ is periodic, then the wave function $\psi_k\left(\vec{r}\right)$ has the form

$$\psi_k\left(\vec{r}\right) = e^{i\vec{k}\cdot\vec{r}} U_k\left(\vec{r}\right)$$
(1.2)

where $U_k\left(\vec{r}\right)$ is periodic in $\vec{r}$ and is known as a Bloch function. Thus the wave function of an electron in a crystal is the product of a periodic lattice function and a plane wave. Periodicity greatly reduces the complexity of solving the Schrödinger equation. The wavevector $\vec{k}$ labeling the wave function serves the same role as the wavevector in the wave function for a free-space electron, and $\hbar\vec{k}$ is known as the crystal momentum [2] of an electron. For some ranges of momentum, the electron velocity is a linear function of its momentum, so the electron in the lattice can be considered as a classical particle; Newton's second law of motion and the law of conservation of momentum determine the trajectory of the electron in response to an external force or a collision.

For a one-dimensional lattice with a period of a, the region $-\pi/a \le k \le \pi/a$ in momentum space is called the first Brillouin zone. The energy band structure is periodic in k with period $2\pi/a$, so it is completely determined by the first unit cell, or the first Brillouin zone in $\vec{k}$ space. In three dimensions, the first Brillouin zone has a complicated shape depending on the specific crystalline structure. *Figure 1-1* shows the Brillouin zone along with several of the most important symmetry points for a face-centered cubic (fcc) lattice, the crystalline structure of semiconductors such as Si, Ge, and GaAs. *Figure 1-2* shows the calculated energy band structures of Si, Ge, and GaAs from the center to the boundary of the first Brillouin zone along two important axes of symmetry, <111> and <100>.

A prominent feature of *Figure 1-2* is the region in which no energy states exist. Such forbidden energy regions exist for all crystalline solids. At a temperature of absolute zero, the lowest energy band that is not fully

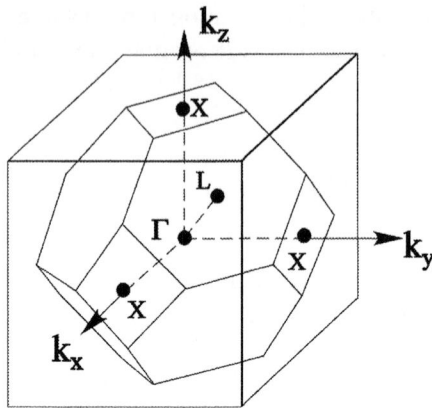

Figure 1-1. The Brillouin zone and several important symmetry points for face-centered cubic lattices such as diamond and zinc blende, the crystalline structures of Si, Ge, and GaAs.

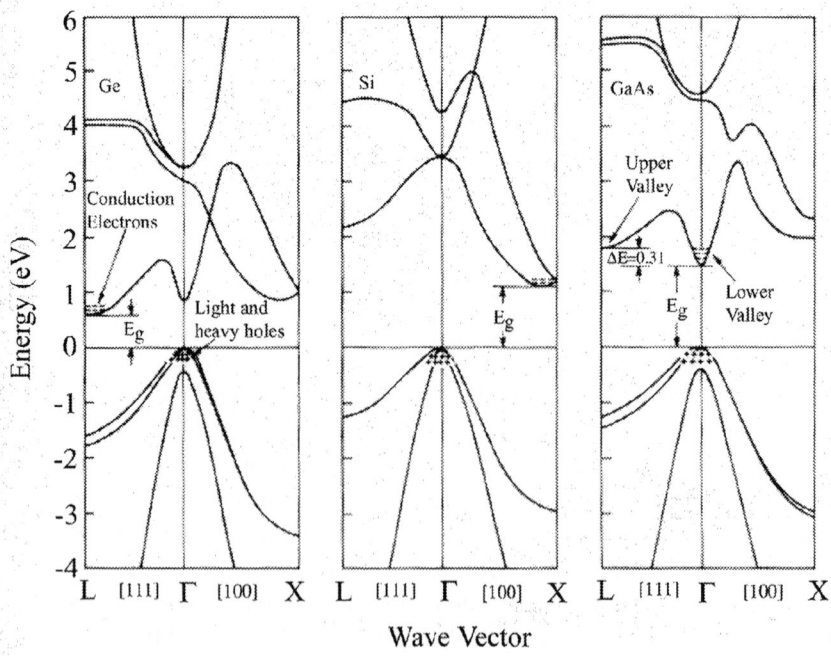

Figure 1-2. Energy band structure along the <100> and <111> axes for Ge, Si, and GaAs in the first Brillouin zone. (Adapted with permission from S. M. Sze, *Physics of Semiconductor Devices*, New York: Wiley, 1981.)

occupied and all higher energy bands are called conduction bands; all lower bands are full of electrons and called valence bands. Whereas the lowest conduction band is partially occupied for conductors, all conduction bands are empty for insulators and semiconductors at absolute zero. The separation between the minimum conduction band and maximum valence band energies is called the bandgap energy, E_g. The main difference between insulators and semiconductors is that the value of the bandgap energy E_g is much larger for insulators than for semiconductors. A semiconductor with its valence band maximum and conduction band minimum at the same wavevector $\bar{k}$ is known as a direct bandgap material (examples include GaAs and InGaAs). A semiconductor with its valence band maximum and conduction band minimum at different wavevectors $\bar{k}$ is known as an indirect bandgap material (examples include Si and Ge).

Electrons in a crystalline material experience external electric fields as well as internal fields generated by other electrons and atoms. When an electric field E_x is applied to a semiconductor, electrons experience a force equal to $|qE_x|$ in a direction opposite to E_x, where q is the unit charge. Whereas an electron in vacuum experiences a constant acceleration and a ballistic trajectory in response to an external force, an electron in a solid also experiences viscous drag forces due to collisions with the lattice. The motion of electrons having energy near the band edges can be described using the effective mass m^*. They attain an average drift velocity $|v_x| = |q\tau E_x/m^*|$, where τ is the average time between collisions. The value of the effective mass is given by

$$m^* = \hbar^2 \left(\frac{d^2 E}{dk^2} \right)^{-1} \tag{1.3}$$

where E is electron energy and $\bar{k}$ is the wavevector. The ratio $q\tau/m^*$ is known as the mobility μ. The average velocity of carriers in a semiconductor due to an applied electric field $\bar{E}$ is the product of the mobility and the electric field:

$$|\vec{v}| = \mu |\vec{E}| \tag{1.4}$$

1.2.2 Carriers in semiconductors

Whereas electrons are the only charge carriers in conductors, both electrons and holes serve as mobile charges in semiconductors. At finite temperature, electrons in the valence band can acquire enough energy to jump into the conduction band, leaving empty states behind in the valence band. When nearly all energy levels in the valence band are occupied by electrons, the empty states are regarded as occupied by positive charges known as holes. The valence states occupied by electrons are considered to be empty of holes.

A perfect semiconductor crystal without any impurities or defects is called an intrinsic semiconductor. In such material, thermal excitation generates electron-hole pairs by providing the energy required for an electron to leave a valence state and enter a conduction state. At thermal equilibrium, the electron and hole concentrations in an intrinsic semiconductor depend on temperature as follows:

$$n = N_C \, e^{-(E_C - E_F)/kT} \tag{1.5}$$

$$p = N_V \, e^{-(E_F - E_V)/kT} \tag{1.6}$$

where k is Boltzmann's constant, T is the absolute temperature, E_F is the Fermi energy, E_C is the minimum energy level of the conduction band, E_V is the maximum energy level of the valence band, and N_C and N_V are the effective densities of states in the conduction band and valence band, respectively. The Fermi energy is defined as the energy level at which electron and hole occupancies are equal; i.e., a state of that energy level is equally likely to be filled or empty. The effective densities of states in conduction band and valence band are defined as

$$N_{C,V} = 2 \left(\frac{2\pi m^* kT}{h^2} \right)^{3/2} \tag{1.7}$$

where h is Planck's constant, and m^* is the density-of-states effective mass (i.e., m_n^* is the effective mass of electrons in the conduction band for N_C, and m_p^* is that of holes in the valence band for N_V). Contours of constant energy in the conduction band of silicon are ellipsoidal rather than spherical, so effective mass is not isotropic. Therefore, to find the effective density of states in the conduction band (N_C), the density-of-states effective mass m_n^* must first be obtained by averaging the effective masses along the appropriate directions:

$$m_n^* = \left(m_l^* \, m_{t1}^* \, m_{t2}^* \right)^{1/3} \tag{1.8}$$

where m_l^*, m_{t1}^*, and m_{t2}^* are the effective masses along the longitudinal and transverse axes of the ellipsoids. As shown in *Figure 1-2*, the two highest valence bands near the center of the first Brillouin zone are approximately parabolic. The band with larger curvature is known as the light hole band due to its smaller effective mass, whereas the other is known as the heavy hole band. The density-of-states effective mass of the valence band averages over both bands according to

$$m_p^* = \left(m_{lh}^{*\,3/2} + m_{hh}^{*\,3/2} \right)^{2/3} \tag{1.9}$$

where m_{lh}^* and m_{hh}^* are the effective masses of the light holes and heavy holes respectively.

Since the numbers of electrons and holes are equal in an intrinsic semiconductor,

$$n = p = n_i \tag{1.10}$$

where n_i is the intrinsic carrier concentration. From equations (1.5), (1.6), and (1.10),

$$n_i = \sqrt{np} = \sqrt{N_C N_V} \, e^{-E_g / 2kT} \tag{1.11}$$

where E_g is the bandgap energy. At a temperature of 300 K, the intrinsic carrier concentration of Si is 1.45×10^{10} cm^{-3} [1].

The carrier concentrations in an intrinsic semiconductor can be dramatically altered by introducing special impurities known as dopants. Introducing dopants known as donors increases the mobile electron concentration, whereas introducing acceptors increases the mobile hole concentration. In doped semiconductors, the concentrations of electrons and holes are no longer equal, and the material is called extrinsic semiconductor. The mass-action law [6] $n_i^2 = np$ still holds: if doping increases the concentration of electrons, the concentration of holes decreases (and vice versa). The predominant carrier is known as the majority carrier and the other is known as the minority carrier. A doped semiconductor is either N type or P type, with the majority carriers being either electrons or holes, respectively.

1.2.3   Optical generation of electron-hole pairs

Photons may be absorbed as they travel through semiconductor material. Sometimes the absorbed energy excites the transition of electrons from the valence band to the conduction band, leaving mobile holes behind. Photoexcited transitions are classified as interband, intraband, or trap-to-band transitions depending on the initial and final energy levels; these are depicted in *Figure 1-3* by the transitions labeled *(a)*, *(b)*, and *(c)*, respectively.

Interband transitions are further classified as either direct or indirect transitions depending on whether an auxiliary phonon is involved. When an electron absorbs a photon of frequency $\omega/2\pi$, its energy increases by $\hbar\omega$. During the absorption, both the total energy and the total momentum of the system must be conserved. The momentum of an incident photon ($\hbar k$) is usually negligible compared to that of the electron. In the absence of additional momentum transfer, an electron transition induced by an incident photon may only occur between energy states with the same wavevector $\bar{k}$. This is known as a direct transition (see *Figure 1-3*).

In silicon, the energy minimum in the conduction band is located at roughly three-quarters of the distance from the Brillouin center along the <100> axis, whereas the energy maximum in the valence band is near the Brillouin zone center. As a result, a photoexcited electron transition requires an additional source of momentum in order to satisfy the conservation of energy and momentum. If the energy of the absorbed photon is the bandgap energy E_g, the momentum required is $3\pi\hbar/4a$, where a is the lattice constant of silicon. This momentum is provided by a quantized lattice vibration called a phonon. A transition involving a phonon is known as an indirect transition (see *Figure 1-3*).

1.3   Silicon-based photodetectors

Photodetectors that transduce optical signals into electronic signals are increasingly important for applications in optical communication and digital photography. Operation of a photodetector comprises: (a) generation of free electron-hole pairs due to impinging light; (b) separation and collection of electrons and holes, possibly with current gain; and (c) production of an output signal through interaction with other components. The main figures of merit for a photodetector are sensitivity to light at the wavelength of interest, response speed, and device noise. In the following sections, several popular silicon-based photosensing devices are discussed: photoconductors, PN and PIN photodiodes, phototransistors, and photogates. Before

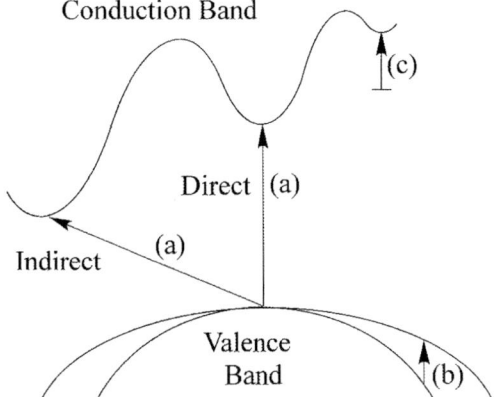

Figure 1-3. Schematic diagram of photoexcitation processes. *(a)* Interband transitions, including a direct transition (with no phonon) and an indirect transition (with a phonon involved). *(b)* Intraband transition. *(c)* Trap-to-band transition.

examining specific devices in detail, several useful concepts will be briefly introduced.

When light impinges on a semiconductor, some fraction of the original optical power is reflected and the rest passes into the material. Inside the solid, interactions between photons and electrons cause a loss of optical power. For a uniform semiconductor material with an absorption coefficient α, the light power at depth x satisfies the relationship

$$P_{ph}(x+dx) - P_{ph}(x) = -\alpha P_{ph}(x)dx \qquad (1.12)$$

Therefore, the optical power traveling through the semiconductor decays exponentially:

$$P_{ph}(x) = P_{ph}(0)e^{-\alpha x} \qquad (1.13)$$

where the surface is taken to be at $x = 0$ and $P_{ph}(0)$ is the optical power at the surface. If the length of the detector along the incident light direction is L, the number of photons absorbed in the detector is

$$N = \frac{P_{ph}(0)}{\hbar\omega}\left(1 - e^{-\alpha L}\right) \qquad (1.14)$$

Figure 1-4 illustrates absorption coefficient as a function of wavelength for several semiconductor materials. As wavelength decreases, the absorption coefficient increases steeply for direct bandgap material like GaAs, whereas absorption coefficients increase more gradually for indirect bandgap material such as Si and Ge. Photons with energy less than the bandgap energy E_g cannot be detected through band-to-band electron transitions, so the maximum (or cutoff) wavelength occurs at

$$\lambda_c = \frac{hc}{E_g} \tag{1.15}$$

for photons with energy equal to E_g, where c is the speed of light. For wavelengths near λ_c, a phonon is required to complete a direct transition in an indirect bandgap material. Hence the probability of absorption decreases significantly for wavelengths near λ_c, as shown in *Figure 1-4* for Si and Ge.

If there is no applied or built-in electric field to separate the photogenerated electron-hole pairs, they will recombine and emit either light or heat. To detect the optical signal, the photogenerated free carriers must be collected. To detect the signal efficiently, the free carriers must be prevented from recombining. The responsivity (R_{ph}) of a photodetector is defined as the ratio of induced current density to optical power density:

$$R_{ph} = \frac{J_{ph}}{P_{ph}'} \tag{1.16}$$

where J_{ph} is the light-induced current density and P_{ph}' is the optical power per unit area of the incident light. The quantum efficiency η is defined as the number of photogenerated carriers per incident photon:

$$\eta = \frac{J_{ph}/q}{P_{ph}'/\hbar\omega} = R_{ph}\frac{\hbar\omega}{q} \tag{1.17}$$

Noise levels determine the smallest optical power that can be detected. The noise sources of a photodetector include shot noise from signal and background currents, flicker noise (also known as $1/f$ noise), and thermal noise from thermal agitation of charge carriers. Noise is often characterized by the noise equivalent power (NEP), defined as the optical power required to provide a signal-to-noise ratio (SNR) of one. Although some authors define the NEP for a 1-Hz bandwidth, the NEP is generally a nonlinear

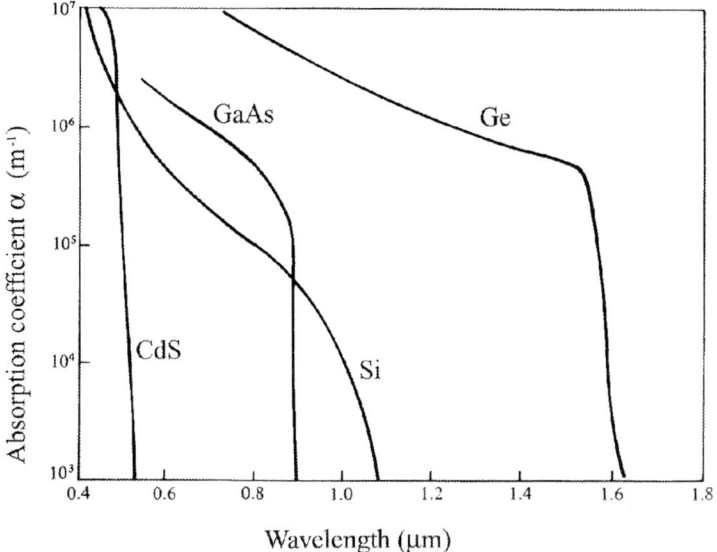

Figure 1-4. Absorption coefficient as a function of wavelength for several semiconductor materials. Below the cutoff wavelength, absorption increases steeply for direct bandgap materials such as GaAs, whereas absorption increases more gradually for indirect bandgap materials such as Si and Ge. (Adapted with permission from J. Wilson and J. F. B. Hawkes, *Optoelectronics: an Introduction,* Englewood Cliffs, New Jersey: Prentice Hall, 1983.)

function of the bandwidth. The reciprocal of the NEP, known as the detectivity *D*, provides an alternative measure; a larger detectivity correlates with improved detector performance. Noise power usually scales with detector area *A* and bandwidth *B*. To normalize for these factors, the specific detectivity D^* is defined as

$$D^* = \frac{\sqrt{AB}}{NEP} \tag{1.18}$$

1.3.1 Photoconductor

The photoconductor is the simplest semiconductor light detector: it consists of a piece of semiconductor with ohmic contacts at both ends. Photons are absorbed as incident light passes through the material, thus generating electrons and holes that increase the conductivity of the semiconductor. Consequently, current flowing through the photoconductor in response to an applied external voltage is a function of the incident optical

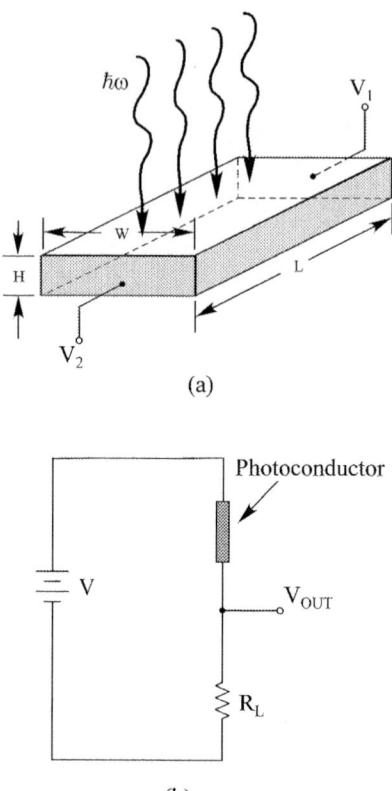

Figure 1-5. (a) Photoconductor structure. (b) Typical photoconductor bias circuit.

power density P_{ph}'. *Figure 1-5* depicts the structure of a photoconductor and a bias circuit for measurement.

Carriers in a semiconductor recombine at the rate $n(t)/\tau$, where $n(t)$ is the carrier concentration and τ is the lifetime of carriers. The excess carrier concentration is the additional carrier concentration relative to the thermal equilibrium concentration, i.e., the "extra" carriers beyond those expected from thermal generation. The photogenerated excess carrier concentration decays exponentially in time after an impulse as $n(t) = n_0 \exp(-t/\tau)$, where the excess carrier concentration at time zero is n_0. For monochromatic light of constant power impinging uniformly on the surface of a photoconductor, the generation rate of electron-hole pairs per unit volume is proportional to the optical power density, whereas the recombination rate depends on the density of photogenerated carriers.

The population of excess carriers increases until the photoconductor reaches equilibrium, i.e., the generation rate balances the recombination rate.

Therefore,

$$R = \frac{\Delta p}{\tau} = \frac{\Delta n}{\tau} = G = \frac{\eta P_{ph}' WL}{\hbar \omega WLH} \tag{1.19}$$

where Δp and Δn are the excess hole and electron densities generated by photon absorption, and L, W, and H are the length, width and height of the photoconductor. These excess carriers increase the conductivity by $\Delta \sigma$, which is

$$\Delta \sigma = q \mu_n \Delta n + q \mu_p \Delta p \tag{1.20}$$

where μ_n is the electron mobility and μ_p is the hole mobility. For an applied voltage of V_{bias}, the photogenerated current density ΔJ is

$$\Delta J = \Delta \sigma \frac{V_{bias}}{L} = \left(\frac{q \mu_n \eta P_{ph}' WL\tau}{\hbar \omega WLH} + \frac{q \mu_p \eta P_{ph}' WL\tau}{\hbar \omega WLH} \right) \frac{V_{bias}}{L}$$

$$= \frac{q \eta P_{ph}' \mu_n}{\hbar \omega H} \left(1 + \frac{\mu_p}{\mu_n} \right) \frac{V_{bias} \tau}{L} \tag{1.21}$$

The light induced current ΔI is the current density times the cross-sectional area:

$$\Delta I = \Delta J W H = \frac{q \eta P_{ph}' W \mu_n}{\hbar \omega} \left(1 + \frac{\mu_p}{\mu_n} \right) \frac{V_{bias} \tau}{L} \tag{1.22}$$

If the primary photocurrent is defined as $I_{ph} \equiv \eta q P_{ph} / \hbar \omega$, where $P_{ph} = P_{ph}' WL$ is the total optical power falling on the detector, then the light-induced current ΔI scales with the primary photocurrent, the carrier lifetime, and bias voltage, and is inversely proportional to length, giving

$$\Delta I = I_{ph} \left(1 + \frac{\mu_p}{\mu_n} \right) \frac{V_{bias} \mu_n \tau}{L^2} = I_{ph} \left(1 + \frac{\mu_p}{\mu_n} \right) \frac{\mu_n E \tau}{L}$$

$$= I_{ph} \left(1 + \frac{\mu_p}{\mu_n} \right) \frac{v_n \tau}{L} = I_{ph} \left(1 + \frac{\mu_p}{\mu_n} \right) \frac{\tau}{t_r} \tag{1.23}$$

where t_r is the transit time of carriers, i.e., the average time required for a carrier to traverse the length of the device. The gain of a photoconductor is defined as the ratio of the light-induced current increment ΔI to the primary photocurrent I_{ph}. The gain G depends on the lifetime of carriers relative to their transit time. For a silicon photoconductor, the gain can exceed 1000, which means that more than one thousand carriers cross the terminals of the photoconductor for each photogenerated electron-hole pair. This simple detector achieves the highest gain of any photo detector—up to 10^6. The response is slow, however, since it depends on the lifetime of carriers. Although it seems counterintuitive that one photogenerated pair produces more than one collected current carrier, the underlying reason is simple: the collection of carriers does not eliminate photogenerated carriers. Only recombination removes them from circulation. Therefore, a photogenerated carrier can travel through the device many times before disappearing through recombination, depending on the ratio τ/t_r.

To analyze the noise performance of the photoconductor, assume that the intensity of the incident optical power is modulated as

$$P(t) = P_{ph}\left(1 + M\,e^{j\omega_0 t}\right) \tag{1.24}$$

where M is the modulation factor and ω_0 is the modulation frequency. The continuity equation

$$\frac{dn(t)}{dt} = -\frac{n(t)}{\tau} + \frac{\eta M P_{ph}\,e^{j\omega_0 t}}{\hbar\omega WLH} \tag{1.25}$$

is solved to find the modulated electron density $n(t)$ in the photoconductor:

$$n(t) = \frac{\tau\eta M P_{ph}\,e^{j\omega_0 t}}{\hbar\omega WLH\left(1 + j\omega_0\tau\right)} \tag{1.26}$$

The modulated hole concentration has the same expression. Therefore, the signal current is given by

$$I(t) = q\left(n(t)\mu_n + p(t)\mu_p\right)\frac{V_{bias}}{L}WH$$

$$= \frac{MI_{ph}\,e^{j\omega_0 t}}{\left(1 + j\omega_0\tau\right)}\left(1 + \frac{\mu_p}{\mu_n}\right)\frac{\tau\mu_n V_{bias}}{L^2} = \frac{MI_{ph}\,e^{j\omega_0 t}}{\left(1 + j\omega_0\tau\right)}\left(1 + \frac{\mu_p}{\mu_n}\right)\frac{\tau}{t_r} \tag{1.27}$$

The mean-square signal power is

$$P_{sig} = \frac{1}{2\left(1+\omega_0^2\tau^2\right)}\left[MI_{ph}\frac{\tau}{t_r}\left(1+\frac{\mu_p}{\mu_n}\right)\right]^2 \tag{1.28}$$

Noise in photoconductive detectors arises mainly from fluctuations in the rates of generation and recombination of electron-hole pairs; this is known as generation-recombination noise. For an intrinsic photoconductor of bandwidth B, the mean-square generation-recombination noise power is given by the following equation [8]:

$$\overline{i_{GR}^2} = \frac{4qBI_0}{\left(1+\omega_0^2\tau^2\right)}\frac{\tau}{t_r} \tag{1.29}$$

where $I_0 = \Delta I$ is the constant current given by equation (1.23). The thermal noise generated by the shunt resistance R_p (from *Figure 1-6*) is

$$\overline{i_{R_p}^2} = \frac{4kTB}{R_p} \tag{1.30}$$

Combining equations (1.28) through (1.30), the signal-to-noise power ratio (*SNR*) of a photoconductor is

$$SNR = \frac{M^2 I_{ph}}{8qB}\left(1+\frac{\mu_p}{\mu_n}\right)\left[1+\frac{kT}{q}\frac{t_r}{\tau}\frac{1}{I_0 R_p}\left(1+\omega_0^2\tau^2\right)\right]^{-1} \tag{1.31}$$

To achieve a specified *SNR*, the average optical power P_{ph} projecting onto the detector must be at least

$$P_{ph}^{min} = \frac{4\hbar\omega B\left(SNR\right)}{\eta M^2\left(1+\frac{\mu_p}{\mu_n}\right)} \times$$

$$\left\{1+\left[1+\frac{M^2 kT}{2q^2 R_p B\left(SNR\right)}\left(\frac{t_r}{\tau}\right)^2\left(1+\omega_0^2\tau^2\right)\right]^{1/2}\right\} \tag{1.32}$$

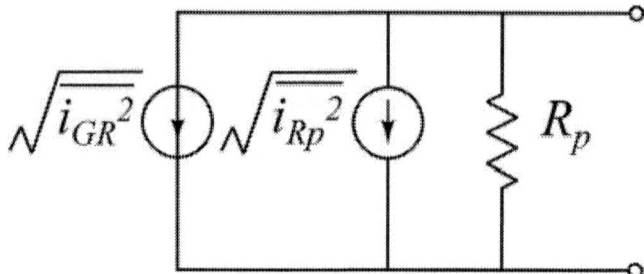

Figure 1-6. Noise equivalent circuit for photoconductor. R_p is the shunt resistance of the semiconductor material, $\overline{i_{GR}^2}$ is the generation-recombination noise, and $\overline{i_{R_p}^2}$ is the thermal noise associated with the resistance R_p.

For an average optical power P_{ph} as in equation (1.24), the root-mean-square (RMS) signal power is $MP_{ph}/\sqrt{2}$, so the NEP may be found by setting $SNR = 1$:

$$NEP = \frac{2\sqrt{2}\hbar\omega B}{\eta M\left(1+\dfrac{\mu_p}{\mu_n}\right)}\left\{1+\left[1+\frac{M^2 kT}{2q^2 BR_p}\left(\frac{t_r}{\tau}\right)^2\left(1+\omega_0^2\tau^2\right)\right]^{1/2}\right\} \quad (1.33)$$

If the photoconductor is extrinsic, I_0 in equation (1.31) is the total current. The conductivity of the material causes a constant current to flow in addition to the photo-induced current, and the generation-recombination noise scales with the total current. The shunt resistance R_p also decreases, so thermal noise increases as well.

The primary advantage of the photoconductor is its high gain, and the primary drawbacks are its slow response and high noise. The following section discusses the characteristics of photodiodes, which provide lower noise and higher speed than photoconductors.

1.3.2   Photodiode

The PN junction photodiode is an important sensor for digital imaging because it is easy to fabricate in bulk silicon complementary metal oxide semiconductor (CMOS) technology, which is inexpensive and widely available. When light irradiates a junction diode, electron-hole pairs are generated everywhere. Electrons and holes generated inside the depletion region will be swept into the adjacent N and P regions, respectively, due to

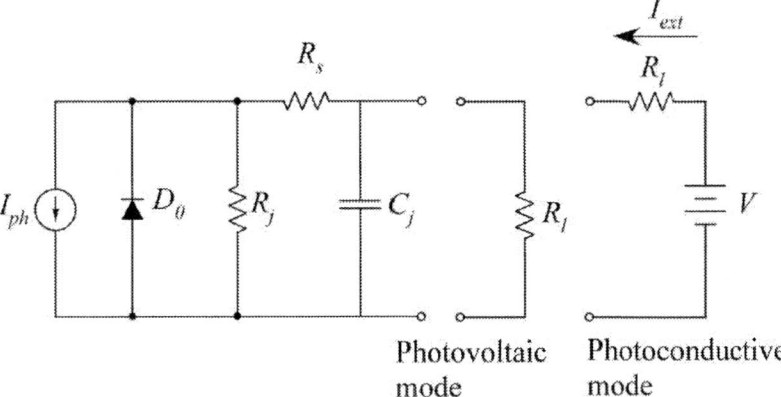

Figure 1-7. Equivalent circuit for a photodiode. I_{ph} is the photocurrent, D_0 is a diode, R_s is the series resistance, R_j is the junction resistance, C_j is the junction capacitance, R_l is the load resistance, and V is the reverse bias voltage.

the electric field across the junction. In addition, electrons and holes generated in the adjacent P and N regions may diffuse into the depletion region and be swept into the other side. Photogenerated carriers swept across the depletion layer may be detected either as photocurrent or as photovoltage. For high quantum efficiency, the depletion layer must be thick in order to absorb as many photons as possible. However, thicker depletion layers increase carrier transit time, resulting in an intrinsic tradeoff between response speed and quantum efficiency.

The photodiode can be operated in two basic modes: photoconductive mode and photovoltaic mode. The equivalent circuit for both modes of operation is shown in *Figure 1-7*, where I_{ph} represents photocurrent, D_0 is a diode, R_s is the series resistance, R_j is the junction resistance, C_j is the junction capacitance, R_l is the load resistance, and V is the reverse bias voltage. If the incident light power is P_{ph}, the photocurrent I_{ph} corresponding to the current source in *Figure 1-7* is

$$I_{ph} = \frac{q\eta P_{ph}}{\hbar\omega} \tag{1.34}$$

where η is the quantum efficiency of the photodiode and ω is the angular frequency of the incident light. In the short-circuit photoconductive mode, the voltage across the photodiode is zero and the external current is I_{ph}. From equations (1.34) and (1.16), the sensitivity of the photodiode is determined by quantum efficiency alone. Quantum efficiency is a function of the absorption coefficient, which is shown in *Figure 1-4* as a function of the wavelength. The absorption coefficient for silicon is weak near the cutoff wavelength λ_c because of the indirect band transition discussed previously. To increase the sensitivity of a vertical photodiode, the junction is usually designed to be very shallow and is reverse-biased in order to widen the depletion region. Most of the impinging light is absorbed in the depletion region when the width of the depletion layer is on the order of $1/\alpha$. For a reverse bias voltage V (from *Figure 1-7*), the current I_{ext} collected by the terminals is

$$I_{ext} = I_{ph} - I_0 \left(e^{\frac{-V + I_{ext}(R_S + R_l)}{U_T}} - 1 \right) + \frac{V - I_{ext}(R_s + R_l)}{R_j} \tag{1.35}$$

where I_0 is the reverse-bias leakage current of the diode D_0 and U_T is the thermal voltage kT/q. The junction resistance R_j is usually large ($\approx 10^8 \, \Omega$), so the third term in equation (1.35) is negligible. In photovoltaic mode, carriers swept through the depletion layer build up a potential across the PN junction until the forward current of the diode balances the photocurrent. For an incident optical power P_{ph}, the resulting open-circuit voltage V_{oc} of the photodiode is

$$V_{oc} = U_T \ln \left(\frac{q \eta P_{ph}}{\hbar \omega I_0} + 1 \right) \tag{1.36}$$

Because of the strong electric field inside the depletion region, the response speed of a photodiode is much faster than that of a photoconductor. Three factors determine the response speed of a photodiode: (a) the diffusion time of carriers outside the depletion layer, (b) the drift time inside the depletion layer, and (c) the time constant due to load resistance and parasitic diode capacitance. To reduce the diffusion time for carriers generated outside the depletion region, the junction should be formed very near the surface. To reduce the drift time for carriers in the depletion region, the junction should be strongly reverse biased so that carriers drift through at the saturation velocity. Strong reverse bias also minimizes parasitic capacitance of the junction diode. Both the width of depletion layer and the carrier drift speed are proportional to $\sqrt{|V_{bias}|}$, which limits the drift-related improvement in response speed.

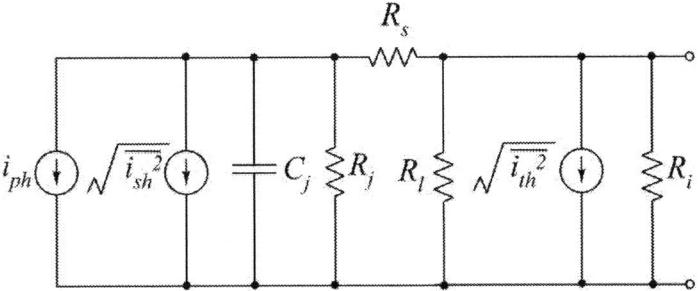

Figure 1-8. Photodiode equivalent circuit for noise analysis: R_s is the series resistance, R_j is the junction resistance, C_j is the junction capacitance, R_l is the load resistance, and R_i is the input resistance of the next stage. The signal photocurrent is i_{ph}; $\overline{i_{sh}^2}$ is the shot noise due to the photocurrent, the background current and the dark current; and $\overline{i_{th}^2}$ is the thermal noise associated with resistances R_s, R_j, R_l, and R_i.

For the analysis of photodiode noise, assume that the signal is modulated as in equation (1.24). The average photocurrent remains the same as in equation (1.34). The root mean square (RMS) optical signal power is $MP_{ph}/\sqrt{2}$ and the corresponding RMS signal current is

$$i_{ph} = \frac{q\eta MP_{ph}}{\hbar\omega\sqrt{2}} \tag{1.37}$$

Figure 1-8 shows the equivalent circuit of a photodiode for noise analysis. Shot noise arises from three sources: (a) the background current I_B generated by ambient background light unrelated to the signal, (b) the dark current I_D resulting from thermal generation of electron-hole pairs inside the depletion region and from reverse saturation current of the diode, and (c) the photocurrent I_{ph}. Each of these currents is generated by an independent random process, and therefore all contribute to the shot noise

$$\overline{i_{sh}^2} = 2q\left(I_{ph} + I_B + I_D\right)B \tag{1.38}$$

where B is the bandwidth of the photodiode. Thermal noise arises from the parasitic resistances of the photodiode. The series resistance R_s is assumed to be negligibly small in comparison to other resistances. Thermal noise is contributed by the junction resistance R_j, the load resistance R_l, and the input resistance R_i (if the photodiode drives another circuit with finite input resistance). These parasitic resistances are modeled by an equivalent resistance R_{eq}, which contributes thermal noise:

$$\overline{i_{th}^{2}} = \frac{4kT}{R_{eq}} B \tag{1.39}$$

Combining equations (1.37) to (1.39), the *SNR* for the photodiode is

$$SNR = \frac{\dfrac{1}{2}\left(\dfrac{q\eta M P_{ph}}{\hbar\omega}\right)^{2}}{2q\left(I_{B}+I_{D}+I_{ph}\right)B+4kT\left(1/R_{eq}\right)B} \tag{1.40}$$

To obtain a specified *SNR*, one must use an optical power P_{ph} equal to or larger than

$$P_{ph}^{min} = \frac{2\hbar\omega B}{\eta M^{2}}(SNR)\left\{1+\left[1+\frac{M^{2}I_{eq}}{qB(SNR)}\right]^{1/2}\right\} \tag{1.41}$$

where the equivalent current I_{eq} is

$$I_{eq} = I_{B}+I_{D}+2kT/qR_{eq} \tag{1.42}$$

For an average optical power P_{ph} as in equation (1.24) and an RMS signal power of $MP_{ph}/\sqrt{2}$, the NEP of the photodiode is

$$NEP = \frac{\sqrt{2}\hbar\omega B}{\eta M}\left\{1+\left[1+\frac{M^{2}I_{eq}}{qB}\right]^{1/2}\right\} \tag{1.43}$$

In the preceding discussion, it has been implicitly assumed that the quantum efficiency η is a known parameter of the device. The remainder of this section describes how the quantum efficiency varies with parameters such as surface reflectivity, absorption coefficient, the width of depletion layer, and the carrier diffusion length.

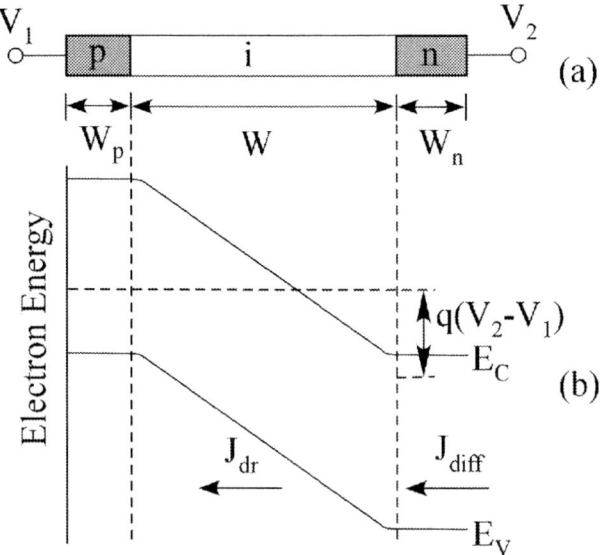

Figure 1-9. (a) Structure and *(b)* energy band diagram for a PIN diode. E_C and E_V correspond to conduction-band energy and valence-band energy respectively.

This model applies to more general PN junction structures such as the PIN diode. The PIN diode is a variant on the simple PN junction in which the P-type and N-type regions are separated by a layer of intrinsic semiconductor. Performance measures such as quantum efficiency and response speed are optimized by choosing the thickness of the intrinsic layer. In a PIN diode, this intrinsic region is fully depleted under reverse bias, so the depletion width is determined primarily by geometry rather than by operating voltage. *Figure 1-9* shows the structure of a PIN photodiode and its energy band diagram under reverse bias. Typically the depletion layer of a PN junction is thin, so photon absorption is assumed to be uniform over the depletion region. In contrast, the PIN diode has a much wider depletion region and absorption is a function of depth in the material.

The total photocurrent consists of a drift current due to carriers generated inside the depletion layer and a diffusion current due to carriers generated outside the depletion region that diffuse into the reverse-biased junction. Therefore, the steady-state current density can be expressed as

$$J_{tot} = J_{dr} + J_{diff} \tag{1.44}$$

If the incident signal at the surface of the PIN photodiode has optical power P_{in} with angular frequency ω, the optical flux within the material is

$$\Phi_0 = \frac{P_{in}(1-R)}{A\hbar\omega} \tag{1.45}$$

where A is the area of the device and R is the reflection coefficient. The generation rate of electron-hole pairs as a function of depth x in the material is

$$G(x) = \alpha\Phi_0 e^{-\alpha x} \tag{1.46}$$

The drift current density and diffusion current density of a P^+N photodiode are derived under the following assumptions: (a) the top P-type layer is much thinner than $1/\alpha$, so current due to carriers generated in the P^+ layer is negligible; (b) recombination within the depletion region is negligible; and (c) thermal generation current is negligible. Under these conditions, the drift current density is

$$J_{dr} = q\int_0^W G(x)dx = q\Phi_0\left(1 - e^{-\alpha W}\right) \tag{1.47}$$

where W is the width of depletion layer. The direction of drift current density J_{dr} is toward the surface of the material. The diffusion current due to carriers generated in the bulk semiconductor near the depletion region is determined by the quasi-static diffusion equation

$$D_p\frac{\partial^2 p_n(x)}{\partial x^2} - \frac{p_n(x) - p_{n0}}{\tau_p} + G(x) = 0 \tag{1.48}$$

where D_p is the diffusion coefficient of holes in the N-type semiconductor, τ_p is the lifetime of the holes, and p_{n0} is the minority hole concentration at thermal equilibrium. Boundary conditions are given by an asymptotic return to equilibrium value $p_{n0}(\infty) = p_{n0}$ and the equilibrium carrier concentration $p_n(W) = p_{n0}\exp(-V/U_T)$ for applied potential V across the junction (V is positive for reverse bias). Equation (1.48) may then be solved to find the distribution of holes in the N-type bulk semiconductor

$$p_n(x) = p_{n0} + \left[p_{n0}\left(e^{\frac{-V}{U_T}} - 1\right) - C_1 e^{-\alpha W}\right]e^{\frac{W-x}{L_p}} + C_1 e^{-\alpha x} \tag{1.49}$$

where $L_p = \sqrt{D_p \tau_p}$ is the diffusion length of holes in the N-type semiconductor, and the coefficient C_1 is

$$C_1 = \frac{\alpha L_p^2}{1 - \alpha^2 L_p^2}\left(\frac{\Phi_0}{D_p}\right) \qquad (1.50)$$

Normally, a relatively large reverse-bias voltage is applied across the PN junction to increase quantum efficiency and response speed. In this case, the boundary condition of hole concentration is approximately $p_n(W) = 0$, and $p_n(x)$ simplifies to

$$p_n(x) = p_{n0} - \left[p_{n0} + C_1 e^{-\alpha W}\right]e^{\frac{W-x}{L_p}} + C_1 e^{-\alpha x} \qquad (1.51)$$

The diffusion current density J_{diff} is given by

$$J_{diff} = -qD_p \left.\frac{\partial p_n}{\partial x}\right|_{x=W} = -\left(qp_{n0}\left(\frac{D_p}{L_p}\right) + q\Phi_0 \frac{\alpha L_p}{1+\alpha L_p}e^{-\alpha W}\right) \qquad (1.52)$$

where the minus sign indicates that current flows toward the surface, which is in the same direction as J_{dr}. The total current density J_{tot} is the sum of drift and diffusion current densities:

$$J_{tot} = J_{dr} + J_{diff} = q\Phi_0 \left(1 - \frac{1}{1+\alpha L_p}e^{-\alpha W}\right) + qp_{n0}\left(\frac{D_p}{L_p}\right) \qquad (1.53)$$

Under normal operating conditions the second term is much smaller than the first, so the current density is proportional to the flux of incident light. Therefore, quantum efficiency is

$$\eta = \frac{J_{tot}/q}{P_{ph}/A\hbar\omega} = \frac{J_{tot}/q}{\Phi_0/(1-R)} = (1-R)\left(1 - \frac{1}{1+\alpha L_p}e^{-\alpha W}\right) \qquad (1.54)$$

To achieve high quantum efficiency, (a) the device must have a small reflection coefficient R and a large absorption coefficient α, and (b) both the diffusion length L_p and the width of depletion layer W must be large in comparison with $1/\alpha$. Response speed degrades when the diffusion current is a large fraction of the total current and when W is large. This imposes an inherent tradeoff between response speed and quantum efficiency for a photodiode.

1.3.3   Phototransistor

In principle, all transistors are light sensitive and may be used as photodetectors. In practice, however, bipolar phototransistors usually exhibit better responsivity because the current flows through a larger volume than in the narrow channel of a field effect transistor. In addition, phototransistors can provide current gain during sensing. A bipolar phototransistor is shown schematically in *Figure 1-10*. In contrast with a conventional PNP transistor, a phototransistor has a large collector-base junction area for collection of photons. Phototransistors usually operate with the base terminal floating. Photogenerated holes in the reverse-biased collector-base junction will be swept into the collector and collected as photocurrent I_{ph}. The emitter-base potential is increased by electrons generated in the base region and swept into the base from the collector. The increase of the emitter-base junction potential causes holes from the emitter to be injected into the base; most of these holes diffuse across the base and appear as additional collector current. Since the base is floating, the emitter current I_E is equal to the collector current I_C:

$$I_C = I_E = I_{CEO} = \left(1 + h_{FE}\right)\left(I_{ph} + I_{eq}\right) \tag{1.55}$$

where h_{FE} is the static common emitter current gain, I_{eq} represents the background current and dark current, and I_{CEO} is the collector-emitter current with open base. The gain of a phototransistor is $1 + h_{FE}$. As with the photodiode, the current across the active junction of a phototransistor contributes to shot noise. With the collector current given in (1.55) and at low temporal frequencies, the output noise power is

$$\overline{i_o^2} = 2qI_C\left(1 + {2h_{fe}^2}\Big/{h_{FE}}\right)B \tag{1.56}$$

where h_{fe} is the small-signal common-emitter current gain (approximately equal to the static gain h_{FE}), and B is the bandwidth of the phototransistor [9]. The net base current is zero, but internally the base current comprises two balanced flows: the junction and recombination currents balance the photocurrent, the dark current, and the background current. Therefore, both

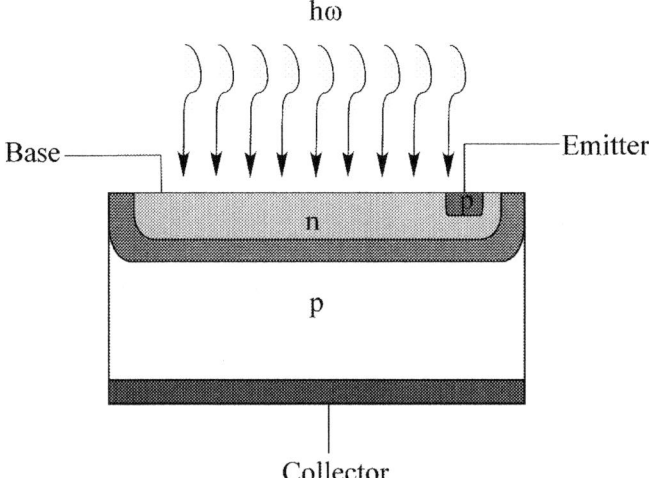

Figure 1-10. Cross-sectional view of a PNP phototransistor. The collector-base junction area is large to increase the collection of photons. Photocurrent biases the floating base region, and the resulting emitter current is $1 + h_{FE}$ times the photocurrent.

balanced components of the base current contribute shot noise, each with spectral density $2q\,(I_C/h_{FE})$. When referred to the output, this appears as $4qh_{fe}^2\,(I_C/h_{FE})$. In addition, the collector current contributes shot noise with spectral density $2qI_C$. Although the signal power is larger because of the phototransistor gain, the noise is larger as well. Given a modulated optical signal as in equation (1.24), the RMS photogenerated current signal is

$$i_{ph} = \left(1 + h_{fe}\right)\frac{q\eta M P_{ph}}{\hbar\omega\sqrt{2}} \tag{1.57}$$

From equations (1.56) and (1.57), the *SNR* is given by

$$SNR = \frac{\dfrac{1}{2}\left(1 + h_{fe}\right)^2\left(\dfrac{q\eta M P_{ph}}{\hbar\omega}\right)^2}{2qI_C\left(1 + \dfrac{2h_{fe}^2}{h_{FE}}\right)B} \tag{1.58}$$

Equation (1.58) may be solved to find the minimum value of the average optical power P_{ph} required to achieve a specified SNR, which is

$$P_{ph}^{min} = \frac{2\hbar\omega B\left(1+h_{FE}\right)\left(1+\dfrac{2h_{fe}^2}{h_{FE}}\right)(SNR)}{\eta M^2\left(1+h_{fe}\right)^2} \times$$

$$\left\{1+\left[1+\frac{M^2\left(1+h_{fe}\right)^2 I_{eq}}{qB(SNR)\left(1+h_{FE}\right)\left(1+\dfrac{2h_{fe}^2}{h_{FE}}\right)}\right]^{\frac{1}{2}}\right\} \tag{1.59}$$

The NEP is given as $MP_{ph}^{min}/\sqrt{2}$ with $SNR = 1$. Under the assumption that $h_{fe} = h_{FE} \gg 1$, the NEP simplifies to

$$NEP = \frac{2\sqrt{2}\hbar\omega B}{\eta M} \times \left\{1+\left[1+\frac{M^2 I_{eq}}{2qB}\right]^{\frac{1}{2}}\right\} \tag{1.60}$$

The tradeoff between low noise and high gain can be adjusted by varying the common-emitter current gain h_{FE}. The response speed of the phototransistor is slow in comparison with the photodiode because of the large depletion capacitance from the large collector-base junction area.

1.3.4 Photogate

The photogate is closely related to the charge-coupled device (CCD), discussed in further detail in the next section. A photogate detector is a metal-oxide-semiconductor (MOS) capacitor with polysilicon as the top terminal. In contrast with the photodetectors discussed above, the photogate transduces optical signals into stored charges rather than voltage or current signals. These stored charges subsequently interact with other components to generate voltage or current signals. The photogate operates by integrating the incident photosignal, so the photogate output is a filtered and sampled version of the incident signal. A cross-sectional view of a photogate and its energy band diagram are shown in *Figure 1-11*. When a positive voltage is applied to the gate above the P-type substrate, holes are pushed away from the surface, leaving a depletion layer consisting of ionized acceptors near the surface. This space charge region causes the energy bands to bend near the semiconductor surface. The potential ψ inside the semiconductor is defined

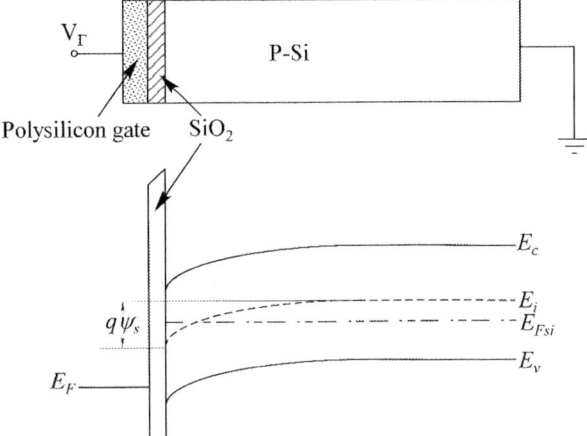

Figure 1-11. Cross-sectional view of the photogate and its energy band structure. The potential ψ is defined as the difference between the intrinsic Fermi-level potential at any location and its value in the bulk semiconductor. The surface potential ψ_s is defined as the value of the potential ψ at the surface of the semiconductor.

as the difference between the intrinsic Fermi-level potential at any location and its value in the bulk semiconductor, and is positive if the energy bands bend downward at the surface. The surface potential ψ_s is defined as the value of the potential ψ at the surface of the semiconductor.

Neglecting potential shifts due to charges at the interface and in the oxide layer and the difference in the work function (the difference in Fermi energy between the polysilicon gate and the semiconductor substrate), the surface potential ψ_s is zero when the applied gate voltage V_G is zero. Applied gate voltage ($V_G \neq 0$) appears partially across the gate oxide as V_{ox} and partially across the depletion region beneath as surface potential ψ_s:

$$V_G = V_{ox} + \psi_s \tag{1.61}$$

Assuming that the substrate is uniformly doped with acceptor density N_A, the charge density on the gate capacitor must balance the ionized charges in the semiconductor; therefore, the potential across the parallel plate capacitor V_{ox} can be expressed as

$$V_{ox} = \frac{qN_A W_D}{C_{ox}} \tag{1.62}$$

where C_{ox} is the oxide capacitance per unit area and W_D is the width of the depletion layer. Poisson's equation

$$\nabla^2 \psi(x) = \frac{qN_A}{\varepsilon_{si}} \tag{1.63}$$

may be solved to find the surface potential ψ_s:

$$\psi_s = \psi(0) - \psi(W_D) = \frac{qN_A W_D^2}{2\varepsilon_{si}} \tag{1.64}$$

Canceling out W_D in (1.62) and (1.64), the gate voltage V_G may be rewritten in terms of ψ_s as

$$V_G = \frac{\sqrt{2qN_A \varepsilon_{si} \psi_s}}{C_{ox}} + \psi_s \tag{1.65}$$

If the photogate collects a signal charge density Q_{sig} at the surface of the semiconductor, the gate voltage V_G becomes

$$V_G = \frac{\sqrt{2qN_A \varepsilon_{si} \psi_s} + |Q_{sig}|}{C_{ox}} + \psi_s \tag{1.66}$$

Rearranging equation (1.66) to find the surface potential ψ_s in terms of the gate voltage and signal charge gives

$$\psi_s = V_G + V_0 - \frac{|Q_{sig}|}{C_{ox}} - \sqrt{2\left(V_G - \frac{|Q_{sig}|}{C_{ox}}\right)V_0 + V_0^2} \tag{1.67}$$

where $V_0 = qN_A \varepsilon_{si}/C_{ox}^2$. Since V_0 is usually very small compared to the gate bias voltage V_G, the surface potential ψ_s is an accurate linear function of the signal charge Q_{sig}.

The derivation of the quantum efficiency of a photogate is similar to that of a photodiode. The depletion region induced by the electric field due to the applied gate voltage is considered to be a shallow junction with junction depth $x_j = 0$; this is consistent with the assumption that the photodiode has no photon absorption in the diffusion region above the junction. The boundary condition for minority-carrier concentration at the edge of the depletion region is

$$n_p\left(W_D\right) = n_{p0}\, e^{-\psi_s/U_T} \tag{1.68}$$

where n_{p0} is the equilibrium concentration of electrons in the P-type substrate. Following the derivation of quantum efficiency for the photodiode and assuming that (a) the substrate thickness is much greater than both the diffusion length L_n and the light penetration length $1/\alpha$ and (b) the depletion width is constant, the quantum efficiency of the photogate is

$$\eta = \left(1 - R\right)\left(1 - \frac{1}{1 + \alpha L_n}\, e^{-\alpha W_D}\right) \tag{1.69}$$

where L_n is the diffusion length of electrons in the P-type substrate. (A more comprehensive derivation is available in van de Wiele [10].) The depletion width depends on the surface potential, which changes as signal charges are collected at the surface, so in the general case quantum efficiency varies over the charge integration interval.

The dominant sources of noise in a photogate include dark current, shot noise, and transfer noise. A photogate sensor usually works in a deep-depletion region while the gate voltage is held at a high voltage in order to attract photogenerated electrons under the gate. During integration, thermally generated minority carriers slowly fill up the potential well at the surface. This thermally generated dark current limits the longest integration time. Thermal generation is a Poisson random process, so the dark current generates a shot noise with variance proportional to its mean value. The arrival of photons is also a Poisson random process, so the photocurrent contributes shot noise as well. As mentioned above, the incident light is transduced into surface charge in a photogate. If the optical power projecting on a photogate is P_{ph}, the signal charge Q_{sig} collected during the integration time t_{int} is

$$\left|Q_{sig}\right| = I_{ph}t_{int} = \frac{q\eta P_{ph}}{\hbar\omega}t_{int} \tag{1.70}$$

where η is the quantum efficiency, $\hbar\omega$ is the photon energy, and I_{ph} is the photocurrent. The shot noise generated by photocurrent I_{ph} and dark current

I_D has a uniform power spectral density (PSD) of $2q(I_{ph} + I_D)$ A^2/Hz. The root-mean-square current noise sampled at time t_{int} is

$$\sqrt{\overline{i_n^2}} = \sqrt{2q\left(I_{ph} + I_D\right)\frac{1}{2t_{int}}} = \sqrt{\frac{q\left(I_{ph} + I_D\right)}{t_{int}}} \tag{1.71}$$

where $1/(2t_{int})$ is the bandwidth of the photogate with sampling rate $1/t_{int}$. The charge fluctuation due to this noise current at the end of the integration period is given by

$$Q_n = \sqrt{q\left(I_{ph} + I_D\right)/t_{int}} \cdot t_{int} = \sqrt{q\left(I_{ph} + I_D\right)t_{int}} \tag{1.72}$$

The *SNR* is obtained by combining equations (1.70) and (1.72):

$$SNR = \left(\frac{Q_{sig}}{Q_n}\right)^2 = \frac{I_{ph}^2 t_{int}^2}{q\left(I_{ph} + I_D\right)t_{int}} = \frac{qt_{int}\left(\dfrac{\eta P_{ph}}{\hbar\omega}\right)^2}{\left(\dfrac{q\eta P_{ph}}{\hbar\omega} + I_D\right)} \tag{1.73}$$

The minimum optical power P_{ph} required to achieve a specified *SNR* is

$$P_{ph}^{min} = \frac{\hbar\omega(SNR)}{2\eta t_{int}}\left\{1 + \left[1 + \frac{4I_D t_{int}}{q(SNR)}\right]^{1/2}\right\} \tag{1.74}$$

The equation above is rearranged with *SNR* = 1 to give

$$NEP = \frac{\hbar\omega}{2\eta t_{int}}\left\{1 + \left[1 + \frac{4I_D t_{int}}{q}\right]^{1/2}\right\} \tag{1.75}$$

The stored charges must transfer at least once to other components to produce an output current or voltage signal, and therefore the conductance of the transfer channel contributes thermal noise. The transfer noise at the output node is composed of this thermal noise integrated over the bandwidth, resulting in *kT/C* noise that is independent of conductance. Consequently, the root-mean-square value of charge fluctuation on the output capacitance C_{EQ} is

$$\sqrt{kTC_{EQ}} \qquad (1.76)$$

The noise increases by this value as it is transferred to the output node.

1.4 Semiconductor image sensors

In previous sections, several devices were discussed as single photodetector elements. Imaging is the process of using arrays of such detectors to create and store images. Whereas the predominant technology for optical imaging applications remains the charge-coupled device (CCD), the active pixel sensor (APS) is quickly gaining popularity. CCD technology has revolutionized the field of digital imaging by enabling diverse applications in consumer electronics, scientific imaging, and computer vision through its high sensitivity, high resolution, large dynamic range, and large array size [11–12]. APS technology is now beginning to enable new applications in digital imaging by offering improved performance relative to CCD technology in the areas of low-power operation, high speed, and ease of integration. The operation and performance of a CCD imager are reviewed, and then the APS imager is introduced. The two technologies are compared to show why APS technology is an attractive alternative to CCD technology. Finally, the technical challenges involved in developing a good APS imager are discussed. More detailed information about APS design is available in chapter 4.

1.4.1 Basic CCD structure and operation

The basic element of CCD technology is similar to the photogate introduced in the previous section. A CCD imager consists of an array of closely spaced MOS capacitors operated in deep depletion on a continuous insulator layer over the semiconductor substrate. A schematic cross-sectional view of a three-phase n-channel CCD using three overlapping polysilicon gates is shown in *Figure 1-12*. Three adjacent polysilicon gates define an imager pixel, with the pixel boundary determined by the voltages applied to the different gates. The gates have the same length as the imager array and are responsible for storing and transferring the signal charges accumulated during the integration phase. Lateral confinement structures isolate signal charges into parallel channels along the direction of charge transfer, and are

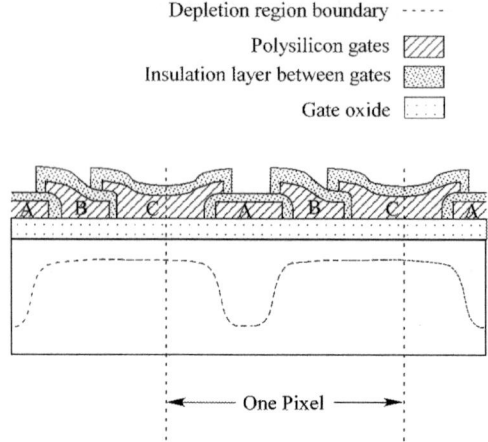

Figure 1-12. Cross-sectional view of a three-phase CCD pixel using three overlapping polysilicon gates (A, B, and C). The pixel boundary is determined by the voltages applied to the gate. The dashed line indicates the depletion boundary during integration, with a high voltage applied to gate A.

typically implemented using a combination of thick field oxide and light doping (see *Figure 1-13*).

During integration, a higher gate voltage is applied to gate A than to gates B and C. This forces the material under gate A into deep depletion, so gate A serves as the signal charge collection and storage element. During the first phase of the three-phase clock, the voltage applied to gate B is pulsed to a high level while the voltage applied to gate A decreases slowly. Thus the charges stored under gate A are transferred into the potential well under gate B. In the second phase, the charges under gate B are again transferred to gate C by pulsing the gate voltage for gate C to a high level while decreasing the voltage on gate B and maintaining a low voltage on gate A. In the third phase, the voltage on gate A returns high while the voltage on gate C decreases. After one cycle of the three-phase clock, the signal charge packet has been transferred to the next pixel. Repeating this process results in a linear motion of the charge packet from the original pixel to the end of the row where it is measured by generating either a voltage or current signal. An optical image represented by the stored charge packets is obtained by scanning through the CCD array.

If every pixel in the CCD imager collects photogenerated charges, the time required to transfer charge packets through the array significantly limits the speed of processing each image frame. One strategy to increase the speed of image readout is to dedicate pixels to storing or shifting out charge

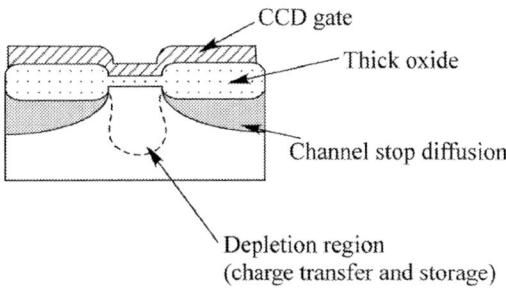

Figure 1-13. Using a combination of techniques including thick field oxide and doping, CCD signal charges are confined laterally in parallel channels along the direction of charge transfer.

packets collected by other pixels. *Figure 1-14(a)* shows a schematic CCD imager using this idea to increase frame rate. However, interleaving storage pixels with charge collecting pixels degrades the resolution of the imager. A frame transfer imager addresses this problem while maintaining fast transfer speed. As shown in *Figure 1-14(b)*, a frame transfer imager uses two CCD arrays of equal size: one records the image and the other stores the image while a single register at the periphery reads out the stored values. The charge packets collected by the imaging area are transferred and temporarily stored in the storage area. During the next cycle of charge accumulation in the imaging array, the signal charges in the storage array are transferred one line at a time to the readout register. The frame transfer imager obtains both high frame rate and fine spatial resolution at the expense of larger chip size.

1.4.2 CCD operating parameters and buried channel CCD

Recording an image using a CCD imager comprises several processes: generation of charges by incident photons; collection of charges by the nearest potential well; transfer of charge packets through the CCD array; and readout by the output preamplifier. The performance of a CCD imager is quantified according to quantum efficiency η, charge collection efficiency η_{CC}, charge transfer efficiency η_{CT}, noise, and response speed.

Quantum efficiency is the number of charges generated per incident photon. CCD imagers can achieve quantum efficiencies as high as 90% using techniques such as thinned high-resistivity substrates, back illumination, and anti-reflection coatings.

Diffusion of photogenerated charges within the substrate restricts the spatial resolution of CCDs. Charges generated at one pixel may diffuse to neighboring pixels, depending on the depth at which the charges are

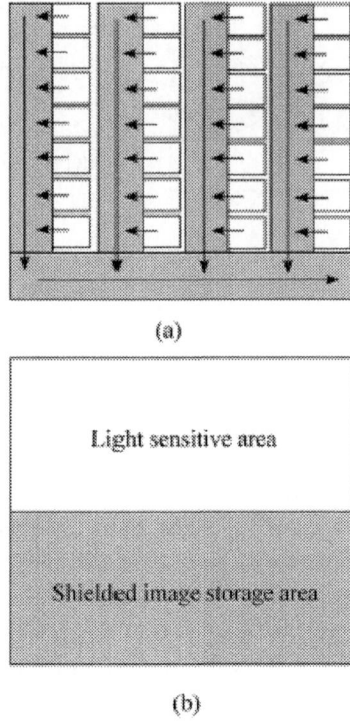

(a)

(b)

Figure 1-14. Two CCD imagers that increase frame rate. *(a)* An imager with shielded pixels (represented by the gray columns) for signal packet transfer interleaved with photosensitive pixels. The bottom row represents a shift register that reads out the signal column-wise. *(b)* An imager with a storage array having the same size as imaging array, known as a frame transfer imager.

generated. Charges generated near the front surface, where the electric field is high, are readily collected by the corresponding potential well. Charges generated deeper in the substrate experience a weak electric field, so those electrons may diffuse into surrounding pixels. This phenomenon is known as a "split event". The diffusing charges may also be lost to trapping and recombination, a phenomenon known as a "partial event". Both types of events cause image blurring and degraded resolution. This is especially important for quantitative applications such as spectroscopy, in which the number of charges collected by a pixel represents the energy of the impinging photon. Charge collection efficiency reflects the degree to which the charges generated by a single photon are collected by a single pixel [13].

CCD imagers rely on the transfer of charge packets from the location where they were initially generated to readout circuitry outside the array. The charge transfer efficiency η_{CT} is defined as the ratio of charge transferred (to the next electrode) to the initial charge stored (under the first electrode). If a packet of total charge Q_0 is transferred n times down to the register, the charge Q_n which reaches the register is

$$Q_n = Q_0 \eta_{CT}^n \tag{1.77}$$

The three basic charge-transfer mechanisms are thermal diffusion, self-induced drift, and fringing field drift. For small charge packets, thermal diffusion is the dominant transfer mechanism. The total charge under the storage gate decreases exponentially with the diffusion time constant τ_{th} [14],

$$\tau_{th} = \frac{4L^2}{\pi^2 D_n} \tag{1.78}$$

where L is the center-to-center space between two adjacent electrodes and D_n is the diffusion coefficient of the carrier. For large charge packets, self-induced drift due to electrostatic repulsion among the charges within the packet is the dominant transfer mechanism. The fringing electric field is independent of the charge intensity, so fringing field drift dominates the charge transfer once most of the charges have shifted.

Three factors determine the charge transfer efficiency: dark current, finite charge transport speed, and interface traps. Dark current arises from thermal charge generation in the depletion region, minority carrier diffusion in the quasi-neutral diffusion region outside the depletion region, and surface recombination current. When a high voltage pulse is applied to an electrode, the surface region immediately transitions into deep depletion. While the gate voltage remains high, thermally generated minority carriers gradually fill the surface potential well, and this artifact corrupts the signal charge packet. Thus, the frequency of the clock must be sufficiently high to minimize the influence of dark current on charge transfer. At very high frequencies, however, the finite charge transport speed degrades the charge transfer efficiency. Short electrode length and high gate voltage help to reduce the charge transfer time. In addition, electrons are used as signal charges rather than holes because of their higher mobility. At intermediate frequencies, interface trapping of the signal charge determines the transfer efficiency. *Figure 1-15* shows trapping and release of carriers from interface traps. As the charge packet enters a potential well, empty traps are filled with signal charges immediately. Some of trapped charges are released quickly and continue on with the correct charge packet as it transfers to the

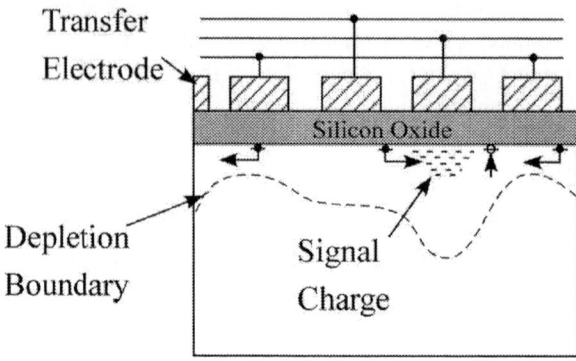

Figure 1-15. Illustration of transfer noise introduced by interface traps.

next electrode. Other interface traps have much slower time constants of release, so the trapped carriers are released into subsequent packets. This delayed release of carriers produces a charge loss from the first packet to the tail of a sequence. Together these mechanisms result in charge transfer inefficiency, which causes signal distortion and phase delay. Interaction of the signal charge packets with interface traps can be decreased significantly by maintaining a constant background charge so that interface traps remain filled. This background charge is called a "fat zero". The drawback of using a fat zero is reduction in dynamic range. Another method for avoiding charge transfer inefficiency due to interface traps is buried channel CCD (BCCD) technology. The transfer inefficiency ($\varepsilon_{CT} = 1 - \eta_{CT}$) of a BCCD is an order of magnitude smaller than that of a surface channel CCD (SCCD) with the same geometry.

The noise floor determines the smallest charge packet that can be detected. It is an important factor in determining the smallest pixel size. Several noise sources limit the performance of a CCD imager. At high signal levels, fixed-pattern noise (FPN) dominates; this noise source arises from pixel-to-pixel variation within the array and can be reduced by adjusting the voltages of the clocks. At intermediate signal levels, shot noise due to the signal current limits the performance. At the lowest signal levels, dark current, fat zero, and amplifier noise limit the performance of the CCD imager.

For further details about CCD devices and imagers, excellent references are available [14–17].

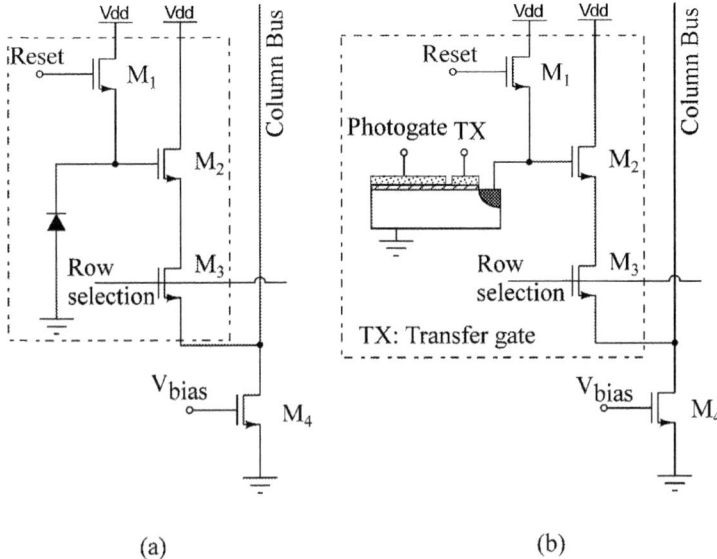

(a)   (b)

Figure 1-16. Two APS photocircuits using *(a)* a photodiode and *(b)* a photogate as the photodetector. The components inside the dashed-line boxes constitute the pixels.

1.4.3   Active pixel sensors

CMOS-based active pixel sensors (APS) have been studied since the 1980s as an alternative to CCD technology. An APS is an imager in which every pixel includes at least one active transistor. Transistors in the APS pixel may operate as both amplifier and buffer in order to isolate the photogenerated charge from the large capacitance of the common output line. While any photodetector may be used, APS pixels often use photodiodes or photogates. *Figure 1-16* illustrates two common photocircuits (circuits inside the pixel) that use photogate and photodiode detectors respectively. APS pixels normally operate in a charge integration mode, but may operate in a continuous mode as well. The operation of different pixels will be discussed further in the following section. The differences between CCD and APS systems will be highlighted here.

Both CCD and APS systems are based on silicon technology, so they have similar sensitivities for visible and infrared wavelengths. CCD imagers require a nearly perfect charge transfer from one electrode to the next. The necessity for nearly perfect charge transfer efficiency becomes clear when the fraction of a charge packet that reaches the output of the CCD imager is examined. A typical three-phase CCD pixel has three electrodes, so a signal packet may be shifted several thousand times on average for an imager with

2048 × 2048 pixels. For a charge transfer efficiency of 99.99%, 26.5% of charges in the original packet will be lost in the array before measurement. Consequently, CCD imagers are (a) radiation soft (i.e., sensitive to displacement damage caused by radiation); (b) difficult to fabricate in large arrays; (c) difficult to integrate with other on-chip electronics; and (d) difficult to operate at very high frequency [18]. Specialized fabrication processes offer improved performance for CCD imagers, but at three times the expense of standard CMOS technology. On the other hand, APS imagers eliminate macroscopic transfer of the charge packet. Thus, charge transfer efficiency is of limited importance, and APS technology avoids the disadvantages associated with maximizing charge transfer efficiency.

Though technically feasible, it is usually impractical to integrate auxiliary functions such as clock drivers, logic control, and signal processing together with the CCD imager. CCD imagers are therefore multi-chip systems, with the resulting large size, heavy weight, and high cost. The primary motivation for developing CMOS-based imagers such as APS is the ease of integration. Virtually all camera electronics such as timing and control signal generators, analog-to-digital converters (ADC), and analog reference generating circuits can be implemented on the same substrate as the imager array. This leads to APS imagers that are "camera-on-a-chip" systems, with digital interfaces that facilitate integration with external systems. As a result, APS imager systems offer compact size, light weight, low cost, and low power consumption. A CMOS APS requires only one-hundredth the power of a CCD system. In addition, APS technology will benefit from the continuous process improvement of mainstream CMOS technology. Other advantages of APS systems include compatibility with transistor-to-transistor logic (TTL) (0–5 V), readout windowing, random access to pixels, and variable integration time.

Despite these advantages and despite twenty years of investigation in APS technology, CCD still remains the dominant imaging technology. Since its inception in 1970, CCD has improved and matured significantly. Using specialized fabrication processes, the integrity of the individual charge packets is maintained as they are physically transferred across the chip to an output amplifier. CCDs provide images with excellent quality for wavelengths ranging from X-rays to infrared radiation. To compete with CCDs, APS technology must offer the same image quality while still retaining its apparent advantages.

APS imagers suffer from fixed pattern noise (FPN) introduced by the mismatch of active transistors in different pixels. This noise can be ameliorated by using the technique of correlated double sampling (CDS). CDS removes systematic offsets by comparing samples taken from the same pixel before and after integration. Typically, there is a CDS circuit for each column, which reduces the FPN for pixels in the same column to an insignificant level. However, column-to-column FPN must be suppressed using other strategies. APS systems also suffer from poor resolution due to a low fill factor. (The fill factor is the ratio of photosensing area in each pixel to the total pixel area.) Typically each pixel uses three active transistors, which limits the fill factor when using standard CMOS processes. Scaling down the feature size improves the fill factor, but at the cost of higher noise and reduced dynamic range due to lower power supplies.

1.5   Information rate

The preceding sections discussed the semiconductor physics underlying phototransduction, several photodetectors, and typical pixel structures for silicon-based imagers. Whereas CCDs are inherently charge-mode devices, APS pixels may use many different methods to convert photogenerated carriers into an electronic output signal. This section will consider the functional performance of candidate APS pixels, as evidenced by their abilities to represent information about incident light intensity.

1.5.1   Information

Imagers are arrays of sensors that transduce optical signals into electronic signals. This transduction is essentially the same as communicating the optical signals while transforming them from one physical variable (light intensity) into another (charge, current, or voltage). Information is a fundamental measure of communication that allows fair comparisons of performance among different technologies and signal representations. The pixel may be considered a communication channel, with the input signal provided by the incident optical signal and the output signal produced by transducing the input signal into an electronic signal. The information rate will be calculated for several different pixels that transduce the optical input signal into an electronic output in the form of charge, current, or voltage.

Information scales logarithmically with the number of possible messages if those messages are equally likely, and reflects the uncertainty in the outcome of the communication—the higher the number of possible outcomes, the more that can be learned from the communication and thus the more information that can be conveyed. The information transmitted through an input/output mapping is known as the mutual information $I(X;Y)$. This is the information that can be learned about the input by observing the output, or vice versa. Mutual information is a measure of dependence between the two signals X and Y. Given the joint probability distribution $p(x, y)$ and the marginal distributions $p(x)$ and $p(y)$, the mutual information rate of signals X and Y is computed as the average uncertainty of the joint distribution relative to the product distribution $p(x)p(y)$:

$$I(X;Y) = \sum_{x,y} p(x,y) \log_2 \frac{p(x,y)}{p(x)p(y)} \tag{1.79}$$

The rate R of information flow at the output is given by the information per sample $I(X;Y)$ times the sample generation rate f_s:

$$R = f_s I(X;Y) \tag{1.80}$$

Information capacity is a fundamental and quantitative bound on the ability of a physical system to communicate information [19]. Information capacity is defined as the maximum mutual information that can be communicated through a channel. This capacity depends only on the physical properties of the channel, such as bandwidth, noise, and constraints on the signal values; it does not depend on the specific details of particular tasks for which the channel may be used. Although task-dependent measures of performance are common in engineering, it is appealing to study the maximum information rate, or channel capacity, especially for sensory devices such as photodetectors that are used for many different tasks.

A Gaussian channel is an additive noise channel in which the noise is a random Gaussian process. For a Gaussian channel, the transmission bandwidth and the ratio of the signal power to the noise power are sufficient to determine the capacity of the channel to transmit information. For a Gaussian channel constrained by an input signal power S, the channel capacity is given by the following relation [20]:

$$C = \Delta f \log_2 \left[1 + \frac{S}{N} \right] \tag{1.81}$$

where S and N are the signal power and noise power, respectively, Δf is the bandwidth of the channel, and the capacity C is in units of bits per second.

In the next few sections, the optical information capacity of APS pixels that transduce the optical signal into charge, current, or voltage will be compared. The transduced representations of the optical signal may be classified as either sampled or continuous. Sampled pixels operate

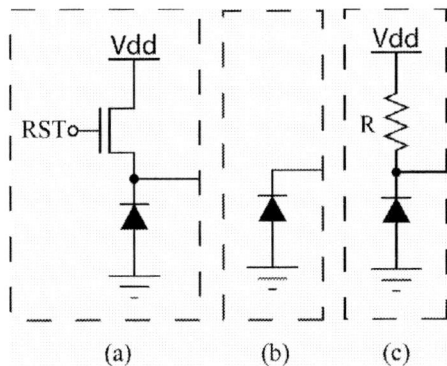

Figure 1-17. Three APS pixels: *(a)* charge-mode (QM) pixel (sampled); *(b)* current-mode (IM) pixel (continuous); and *(c)* voltage-mode (VM) pixel (continuous).

essentially by integrating the photocurrent onto a capacitor during a sampling period, then resetting the capacitor, and then integrating again during the next sampling period. In contrast, continuous pixels transduce the photocurrent waveform continuously into output current or voltage signals. The three APS pixel structures shown in *Figure 1-17* are considered here: charge-mode pixel (sampled), current-mode pixel (continuous), and voltage-mode pixel (continuous).

Each of these pixels operates together with a readout circuit, as indicated in *Figure 1-18*. The readout circuitry inevitably contributes additional readout noise to the signal of interest. *Figure 1-18(a)* shows a general architecture that places all readout circuitry at the column level. As indicated in *Figure 1-18(b)*, sometimes this readout circuitry is split into the pixel level and the column level. The latter circuit shows a transistor within the pixel operating as a source follower, with the current source shared over the entire column. In order to focus on the essential characteristics of the different pixel structures under study, the noise contributions of the readout circuits are neglected in the following discussion.

To compute the information rates for these three pixel structures, assume that the input is an intensity modulated optical signal with average photocurrent I_{ph} and total signal power $\sigma_s^2 I_{ph}^2$. The variable I_B accounts for any constant current through the photodiode that is unrelated to the incident optical signal, such as dark current and current due to background light. The contrast power of the incident photosignal is the variance σ_s^2 of the normalized photosignal (i.e., the optical signal normalized by its mean value). The normalized photosignal is independent of the mean illumination level and shows variations due to scene details, such as the relative reflectivity of objects within a scene [21–22]. For a sinusoidally modulated signal as in equation (1.24), the contrast power is $\sigma_s^2 = M^2/2$.

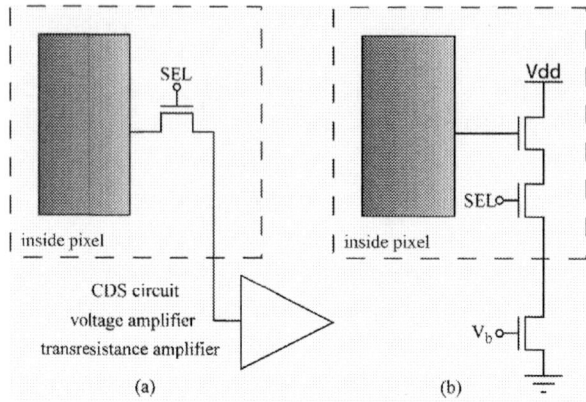

Figure 1-18. Schematics of imager pixels (inside the dashed lines) and the corresponding circuitry for two readout strategies. *(a)* All readout circuitry is at the column level. *(b)* The readout circuitry is split into pixel-level and column-level components.

1.5.2   Charge-mode pixels

The charge-mode pixel is the basic unit of the CMOS active pixel sensor (APS). The charge-mode pixel shown in *Figure 1-17(a)* consists of a photodiode and a reset transistor. In an imager, these elements are augmented by a source follower and a switch for row selection within the pixel. A current source for the source follower and a correlated double sampling (CDS) readout circuit are shared by pixels in the same column (as discussed in the previous section). The operation of the charge-mode pixel has three stages: reset, integration and readout. In the reset stage, the reset transistor is turned on and the detecting node is initialized to a high value. During integration, the photocurrent collected by the photodiode is accumulated by discharging the detecting node. After the integration period, the pixel value is read out by the CDS circuit and stored at the column level. The cycle starts again by resetting the pixel and beginning to accumulate photocurrent. The stored values in each column are read out before the end of the next cycle.

The input-referred noise may be calculated separately for each reset and integration cycle and then summed to determine the total input-referred noise (neglecting readout noise as discussed above).

The reset period is normally much shorter than the settling time for the active pixel sensor; therefore, steady-state operation is not achieved and the reset noise power must be calculated using temporal analysis as in Tian et al. [23]. The mean-square noise voltage at the end of the reset period is

$$\overline{V_n^2(t_r)} = \frac{1}{2}\frac{kT}{C_{out}}\left(1 - \frac{t_{th}}{\left(t_r - t_1 - t_{th}\right)^2}\right) \qquad (1.82)$$

where t_r is the reset time, t_1 is the time that the reset transistor operates in the above threshold region, and t_{th} is the time required to charge the detecting node capacitance C_{out} up to the thermal voltage (using the current of the reset transistor at the point where it enters subthreshold operation). Since the reset time is usually several microseconds and therefore three orders of magnitude larger than t_1 and t_{th}, the mean-square noise voltage is approximately equal to $kT/(2C_{out})$.

During integration, the reset transistor is turned off. The only noise source is the shot noise from the photodiode. If it is assumed that the capacitance at the detecting node remains constant during integration, the mean-square noise voltage sampled at the end of integration will be

$$\overline{V_n^2(t_{int})} = \frac{q\left(I_{ph} + I_B\right)}{C_{out}^2} t_{int} \qquad (1.83)$$

where t_{int} is the integration time.

Thus, the information rate of the charge-mode pixel is given by

$$I_{QM} = \frac{1}{2t_{int}}\log_2\left[1 + \frac{\dfrac{\sigma_s^2 I_{ph}^2}{C_{out}^2}t_{int}^2}{\dfrac{kT}{2C_{out}} + \dfrac{q\left(I_{ph} + I_B\right)t_{int}}{C_{out}^2}}\right] \qquad (1.84)$$

where $1/(2t_{int})$ is the bandwidth of the pixel with sampling rate $1/t_{int}$. The reset time t_r can be neglected since it is normally much shorter than the integration time t_{int}. Thus, the information rate of the charge-mode pixel is a function of the integration time t_{int}, the average photocurrent I_{ph}, the capacitance of the detecting node C_{out}, the absolute temperature T, and the background current I_B. If $1/(2t_{int})$ is considered to be the bandwidth Δf of the charge-mode pixel, equation (1.84) may be rewritten as

$$I_{QM} = \Delta f \log_2\left[1 + \frac{\sigma_s^2 I_{ph}^2}{\left(2kTC_{out}\Delta f + 2q\left(I_{ph} + I_B\right)\right)\Delta f}\right] \qquad (1.85)$$

1.5.3   Current-mode pixels

The current-mode pixel shown in *Figure 1-17(b)* directly transduces the photocurrent generated by the photodiode into a current signal. The pixel consists solely of a photodiode and serves as a point of comparison with the continuous voltage-mode pixel and the sampled charge-mode pixel. In practice, this photodiode is used with a switch for row selection within the pixel, and possibly with some other current steering or amplification components as well.

Since the photodiode contributes shot noise of $2q(I_{ph} + I_B)\Delta f$ and the active load circuitry (not shown in *Figure 1-17(b)*) also contributes a variance of $2q(I_{ph} + I_B)\Delta f$, the mean-square noise current is

$$\overline{I_n^2} = 4q\left(I_{ph} + I_B\right)\Delta f \tag{1.86}$$

where Δf is the bandwidth of the measurement. Thus, the information rate of the current-mode pixel is given by

$$I_{IM} = \Delta f \log_2\left[1 + \frac{\sigma_s^2 I_{ph}^2}{4q\left(I_{ph} + I_B\right)\Delta f}\right] \tag{1.87}$$

which is a function of the bandwidth Δf, the average photocurrent I_{ph}, and the background current I_B. If the currents (I_{ph} and I_B) and contrast power (σ_s^2) are fixed, the information rate is a monotonic function of frequency bandwidth and approaches its maximum value as the frequency bandwidth increases without bound:

$$
\begin{aligned}
I_{IM}^{max} &= \lim_{\Delta f \to \infty} \Delta f \log_2\left[1 + \frac{\sigma_s^2 I_{ph}^2 / 4q\left(I_{ph} + I_B\right)}{\Delta f}\right] \\
&= \frac{\sigma_s^2 I_{ph}^2 / 4q\left(I_{ph} + I_B\right)}{\ln 2}
\end{aligned} \tag{1.88}
$$

The current-mode pixel shown in *Figure 1-17(b)* is an idealization intended to illustrate the performance associated with direct measurement of a photocurrent. Some active load is clearly required to communicate the photocurrent to the outside world. One implementation for such an active load is a source-follower transistor, as shown in *Figure 1-19*. In this case, the

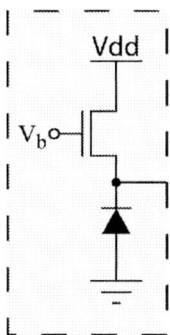

Figure 1-19. Logarithmic voltage-mode pixel (continuous).

pixel produces an output voltage that is logarithmically related to the input photocurrent. The input photocurrent and the shot noise are identical to that for the current-mode pixel described above, and both are transferred into the signal and noise components of the voltage output. The transfer functions for signal and noise are identical, so the information rate for the logarithmic voltage-mode pixel is the same as that given in equation (1.87) for the current-mode pixel.

1.5.4   Voltage-mode pixels

The linear voltage-mode pixel shown in *Figure 1-17(c)* converts the photocurrent into an output voltage signal using a linear resistor. The mean-square signal voltage is given by

$$\overline{V_{sig}^2} = \sigma_s^2 I_{ph}^2 R^2 \tag{1.89}$$

Since the photodiode contributes a variance of $2q\,(I_{ph} + I_B)\,R^2\,\Delta f$ and the resistance contributes a variance of $4kTR\Delta f$, the mean-square noise voltage is given by

$$\overline{V_n^2} = \left[4kTR + 2q\left(I_{ph} + I_B \right) R^2 \right]\Delta f \tag{1.90}$$

Thus, the information rate of the linear voltage-mode pixel is given by

$$I_{VM} = \Delta f \log_2 \left[1 + \frac{\sigma_s^2 I_{ph}^2 R^2}{\left(4kTR + 2q\left(I_{ph} + I_B \right) R^2 \right)\Delta f} \right] \tag{1.91}$$

The information rate of the linear voltage-mode pixel given in (1.91) is a monotonic function of bandwidth Δf when other variables are fixed, and approaches a maximum value as the bandwidth increases without bound:

$$
\begin{aligned}
I_{VM}^{max} &= \lim_{\Delta f \to \infty} \Delta f \log_2 \left[1 + \frac{\sigma_s^2 I_{ph}^2 R^2}{\left(R + 2q\left(I_{ph} + I_B \right) R^2 \right) \Delta f} \right] \\
&= \frac{1}{\ln 2} \frac{\sigma_s^2 I_{ph}^2}{4kT / R + 2q\left(I_{ph} + I_B \right)}
\end{aligned}
\tag{1.92}
$$

1.5.5   Comparison

Unlike the current-mode and linear voltage-mode pixels, the information rate of the charge-mode pixel is not a monotonic function of the measurement bandwidth Δf. *Figure 1-20* shows the information rate of the charge-mode pixel as integration time varies; the information exhibits a maximum at a finite integration time. This result apparently contradicts the intuitive idea that high-quality images require long integration times (under the implicit assumption that the scene is static during the integration time). However, information rate reflects both the quality of the picture and the temporal variation of the scene. Integration times that are too short result in poor signal-to-noise ratios, whereas integration times that are too long sacrifice details regarding changes in the scene. In contrast, information rates for the current-mode and voltage-mode pixels increase monotonically as the bandwidth increases to infinity.

The information rate also varies with other parameters of the pixels. The information rate of a charge-mode pixel decreases with increasing capacitance at the detecting node. The information rate of the linear voltage-mode pixel decreases with decreasing resistance according to equation (1.91). Increasing temperature reduces the information rate for both charge-mode and voltage-mode pixels, and increasing background current I_B (including dark current and photocurrent unrelated to the signal) reduces the information rate for all pixels.

In *Figures 1-21* and *1-22*, the information rates of the three pixels are compared (a) for a fixed bandwidth of $1/(2 \times 30 \text{ ms}) = 16.7 \text{ Hz}$ as the photocurrent varies, and (b) for an optimal bandwidth as the photocurrent varies (with limiting values of detecting node capacitance for the charge-mode pixel and of resistance for the linear voltage-mode pixel). The first case corresponds to an integration time of 30 milliseconds, which is typical

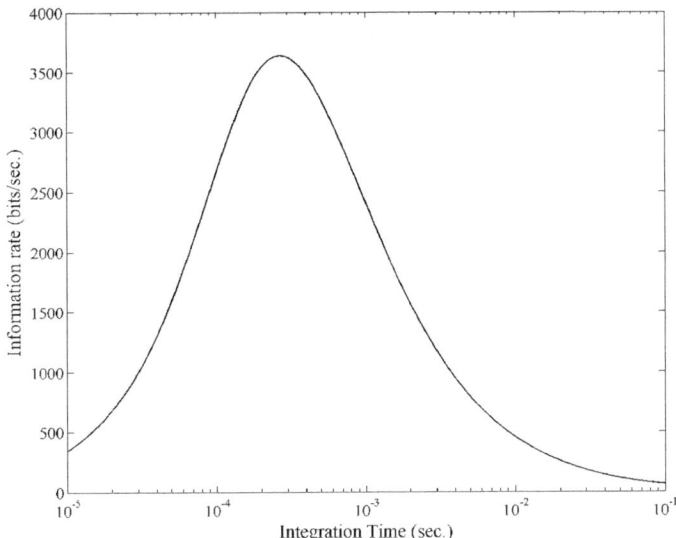

Figure 1-20. The information rate of a charge-mode pixel as a function of integration time t_{int} for an average photocurrent of 100 fA, a background current of 2 fA, a detecting node capacitance of 10 fF, a temperature of 300 K, and a contrast power of 0.1.

for video applications. The second case illustrates the limiting behavior as the photocurrent varies.

The first comparison is of the information rates for fixed measurement bandwidth and integration time, as given by equations (1.85), (1.87), and (1.91). In this case, each of the pixels has two distinct regimes of behavior as a function of photocurrent.

At sufficiently low values of the photocurrent (e.g., $I_{ph} \ll I_B$), the I_{ph} term in the denominator is negligible compared to the other terms, and the information rates of the three pixels can be approximated as

$$I = \Delta f \log_2 \left[1 + \frac{\sigma_s^2 I_{ph}^2}{a \Delta f} \right] \qquad (1.93)$$

where a is independent of photocurrent and takes different values for the three pixel structures under study: $a_{QM} = 2kTC_{out}\Delta f + 2qI_B$, $a_{IM} = 4qI_B$, and $a_{VM} = 4kT/R + 2qI_B$.

For sufficiently high photocurrents, the term I_{ph} dominates the other terms in the denominator, and the information rates of the three pixels can be approximated as

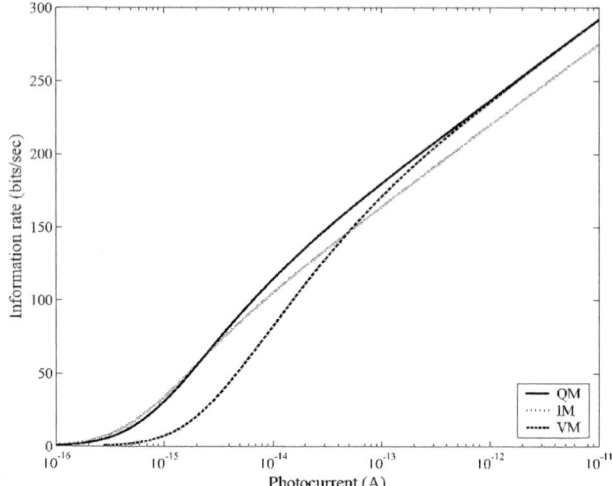

Figure 1-21. Information rates of the charge-mode pixel (QM), the current-mode pixel (IM), and the linear voltage-mode pixel (VM) as functions of the photocurrent for a measurement bandwidth of 16.7 Hz. Pixel background current is 2 fA, temperature is 300 K, contrast power is 0.1, detecting node capacitance is 10 fF, and resistance is 300 GΩ.

$$I = \Delta f \log_2 \left[\frac{\sigma_s^2 I_{ph}}{b\Delta f} \right] \qquad (1.94)$$

where b is independent of photocurrent and takes different values for the three pixel structures under study: $b_{QM} = 2q$, $b_{IM} = 4q$, and $b_{VM} = 2q$. At very low values of the photocurrent, information increases quadratically with photocurrent. For intermediate photocurrents, information increases in proportion to $\log(I_{ph}^2) = 2\log(I_{ph})$; for large photocurrents, it increases as $\log(I_{ph})$. These properties of the information rates for the three pixel structures are shown in the S-shaped curves of *Figure 1-21*. For very low photocurrents I_{ph} and for small I_B, the information rate of the current-mode pixel is the largest of the three. For large photocurrents, the current-mode pixel has the smallest information rate, because the shot noise contributed by the active load of this pixel is (in this case) larger than both the reset noise of the charge-mode pixel and the thermal noise of the voltage-mode pixel.

The second comparison is of idealized pixels at limiting values of detecting node capacitance and load resistance, giving zero reset noise and zero thermal noise for the charge-mode pixel and the linear voltage-mode pixel, respectively. Under these conditions, the information rates of all three

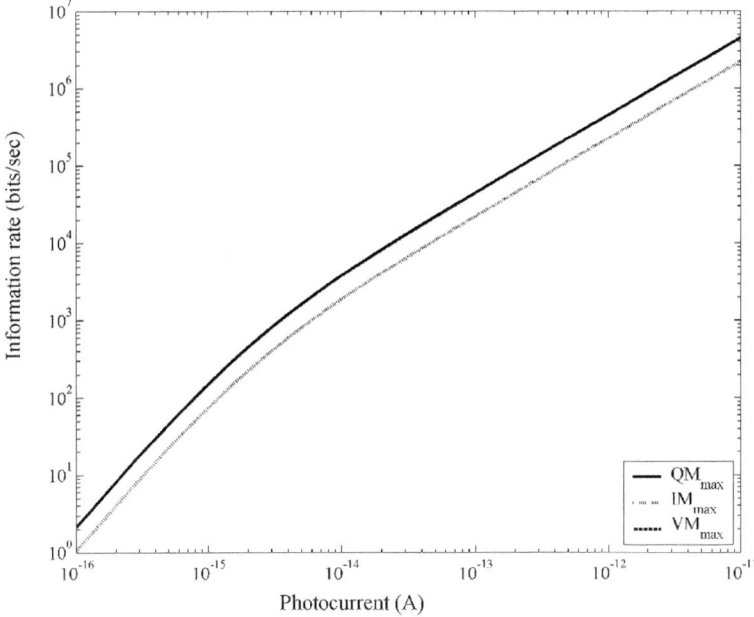

Figure 1-22. The maximum information rates of charge-mode pixel (QM), current-mode pixel (IM), and linear voltage-mode pixel (VM_R) as functions of the photocurrent.

pixels increase monotonically as measurement bandwidth increases. *Figure 1-22* shows the information rates for infinite bandwidth as a function of photocurrent. This figure shows that the information rates for the charge-mode pixel and the voltage-mode pixel are identical when reset noise and thermal noise are neglected. In addition, the optimal information rate increases as I_{ph}^2 for small photocurrents and as I_{ph} for large photocurrents.

1.6 Summary

To understand semiconductor phototransduction, relevant concepts from semiconductor physics have been reviewed. As an indirect bandgap material, silicon is not an ideal material for photodetection because of its low absorption coefficient. However, the maturity of silicon fabrication technology and silicon integrated circuitry development makes silicon the most popular material for image sensors.

Several photodetectors relevant to digital photography were examined with respect to performance metrics such as responsivity, noise, and response time. The response speeds of all the photodetectors discussed here are adequate for imaging applications; therefore, the primary focus was on their relative quantum efficiency and NEP, which reflect responsivity and device noise, respectively. Photoconductors and phototransistors have gains much larger than one, and consequently have higher noise than other photodetectors such as photodiodes and photogates. In addition, the conductivity of on-chip resistors is subject to significant process variation, so photoconductors are rarely used in image sensors. Choices among the remaining three photodetectors depend on factors such as light intensity, imaging frame rate, and pixel size.

Photodiodes and photogates are compatible with standard CMOS fabrication technology, so they are the most popular choices for APS image sensors. Although CCD image sensors are most common in digital photography, APS image sensors are gaining popularity because of their low voltage operation, low power consumption, highly compact size, and low cost. The information rates of APS pixels were compared by considering them as Gaussian channels through which optical signals are transduced and communicated. All pixels show similar information rates at 30 frames per second, and show similar trends in the information capacity as the photocurrent varies.

BIBLIOGRAPHY

[1]   S. M. Sze, *Physics of Semiconductor Devices,* 2nd ed., New York: Wiley, 1981.

[2]   Charles Kittel, *Introduction to Solid State Physics,* 7th ed., New York: Wiley, 1996.

[3]   Wallace B. Leigh, *Devices for Optoelectronics,* New York: Marcel Dekker, 1996.

[4]   Jasprit Singh, *Optoelectronics: an Introduction to Materials and Devices,* New York: McGraw-Hill, 1996.

[5]   David K. Ferry, *Quantum Mechanics: an Introduction for Device Physicists and Electrical Engineers,* 2nd ed., Philadelphia: Institute of Physics Publishing, 2001.

[6]   R. Muller and T. Kamins, *Device Electronics for Integrated Circuits,* 2nd ed., New York: Wiley, 1986.

[7]   J. Wilson and J. F. B. Hawkes, *Optoelectronics: an Introduction,* Englewood Cliffs, NJ: Prentice Hall, 1983.

[8]   A. van der Ziel, *Fluctuation Phenomena in Semiconductors,* New York: Academic Press, 1959.

[9]   H. M. Franciso, R. C. Eugene, and A. van der Ziel, "Noise in phototransistors," *IEEE Trans. Electron Devices,* vol. ED-18, no. 6, pp. 340–346, 1971.

[10] F. van de Wiele, "Photodiode quantum efficiency," in *Solid State Imaging,* P. G. Jespers, F. van de Wiele, and M. H. White, Eds. Leyden: Noordhoff, 1976, pp. 47–90.

[11] M. Lesser and P. Vu, "Processing of back illuminated 4096 × 4096 Fairchild CCDs at the University of Arizona," *Proc. SPIE,* vol. 4306, pp. 196–204, 2001.

[12] P. Suni, V. Tsai, and P. Vutz, "4K × 2K pixel three-side buttable scientific CCD imager design and fabrication," *Proc. SPIE,* vol. 2172, pp. 190–207, 1994.

[13] J. Janesick, K. Klaasen, and T. Elliott, "CCD charge collection efficiency and the photon transfer technique," *Proc. SPIE: Solid State Imaging Arrays,* vol. 570, pp. 7–19, 1985.

[14] C. K. Kim, "The physics of charge-coupled devices," in *Charge-Coupled Devices and Systems,* M. J. Howes and D. V. Morgan, Eds. Wiley: New York, 1979.

[15] C. Sequin and M. Tompsett, *Charge Transfer Devices*, Academic Press: New York, 1975.

[16] J. D. E. Beynon and D. R. Lamb, *Charge-Coupled Devices and their Applications*, New York: McGraw-Hill, 1980.

[17] Albert J. P. Theuwissen, *Solid-State Imaging with Charge-Coupled Devices*, Boston: Kluwer, 1995.

[18] E. R. Fossum, "Active pixel sensors: are CCDs dinosaurs?" *Proc. SPIE*, vol. 1900, pp. 2–14, 1993.

[19] C. E. Shannon, "A mathematical theory of communication," *Bell Syst. Tech. J.*, vol. 27, pp. 379–423 and 623–656, 1948.

[20] C. E. Shannon, "Communication in the presence of noise," *Proc. Inst. Radio Eng.*, vol. 37, pp. 10–21, Jan. 1949.

[21] Whitman A. Richards, "Lightness scale from image intensity distributions," *Appl. Opt.*, vol. 21, no. 14, pp. 2569–2582, 1982.

[22] Simon B. Laughlin, "A simple coding procedure enhances a neuron's information capacity," *Z. Naturforsch., C: Biosci.*, vol. 39, no. 9–10, pp. 910–912, 1981.

[23] Hui Tian, Boyd Fowler, and Abbas El Gamal, "Analysis of temporal noise in CMOS photodiode active pixel sensor," *IEEE J. Solid-State Circuits*, vol. 36, no. 1, pp. 92–101, Jan. 2001.

Chapter 2

CMOS APS MTF MODELING

Igor Shcherback and Orly Yadid-Pecht
The VLSI Systems Center
Ben-Gurion University
P.O.B. 653 Beer-Sheva 84105, ISRAEL

Abstract: The modulation transfer function (MTF) of an optical or electro-optical device is one of the most significant factors determining the image quality. Unfortunately, characterization of the MTF of the semiconductor-based focal plane arrays (FPA) has typically been one of the more difficult and error-prone performance testing procedures. Based on a thorough analysis of experimental data, a unified model has been developed for estimation of the overall CMOS active pixel sensor (APS) MTF for scalable CMOS technologies. The model covers the physical diffusion effect together with the influence of the pixel active area geometrical shape. Agreement is excellent between the results predicted by the model and the MTF calculated from the point spread function (PSF) measurements of an actual pixel. This fit confirms the hypothesis that the active area shape and the photocarrier diffusion effect are the determining factors of the overall CMOS APS MTF behavior, thus allowing the extraction of the minority-carrier diffusion length. Section 2.2 presents the details of the experimental measurements and the data acquisition method. Section 2.3 describes the physical analysis performed on the acquired data, including the fitting of the data and the relevant parameter derivation methods. Section 2.4 presents a computer model that empirically produces the PSF of the pixel. The comparisons between the modeled data and the actual scanned results are discussed in Section 2.5. Section 2.6 summarizes the chapter.

Key words: CMOS image sensor, active pixel sensor (APS), modulation transfer function (MTF), point spread function (PSF), diffusion process, parameter estimation, modeling.

2.1   Introduction

Recent advances in CMOS technology have made smaller pixel design possible. A smaller pixel can improve the spatial resolution of an imager by increasing its pixel count. However, the imager is then more susceptible to carrier crosstalk, which works against the spatial resolution. In addition, smaller pixels tend to have an inferior signal-to-noise ratio (SNR) because the photon flux that they receive is reduced. Charge collection and pixel spatial resolution analysis are therefore important in designing a smaller pixel.

The modulation transfer function (MTF) is the most widely used spatial resolution index. It is defined as the transfer ratio between the imager input and output signal modulations as a function of the spatial frequency of the input signal. Several MTF models have been developed for CCD imagers. Numerous factors affecting MTF have been discussed, including carrier diffusion, epi-layer thickness, substrate doping, and others [1–9]. However, these models do not work for CMOS imagers very well. Most CMOS imagers use photodiodes for charge collection, and the collection mechanism of photodiodes differs from that of the potential well used by CCDs. Only a portion of the charges is collected in the region enclosed by the photodiode. Considerable numbers of charge carriers are generated in the photodiode surroundings, and because of the smaller fill factor, these can be collected by any nearby diode. An applicable model for charge collection MTF is needed for efficient designs.

This chapter provides a logical extension of a pixel response analysis by Yadid-Pecht [10]. The pixel photoresponse is analyzed and a comprehensive MTF model is described, enabling a reliable estimate of the degradation of the imager performance.

2.1.1   Optical and modulation transfer functions

The optical transfer function (OTF) determines the output for any given input. It is defined in a way similar to the Fourier transform of the spread function in electronics; i.e., it is the "impulse" response normalized to its own maximum value, which occurs at zero spatial frequency:

$$OTF = \tau\left(\omega_x, \omega_y\right) = \frac{\iint_\infty s\left(x', y'\right)\exp\left[-j\left(\omega_x x' + \omega_y y'\right)\right]dx'dy'}{\iint_\infty s\left(x', y'\right)dx'dy'}$$

$$= \frac{S\left(\omega_x, \omega_y\right)}{S\left(0, 0\right)} = MTF * \exp\left[jPTF\right]$$

$$(2.1)$$

where $S(\omega_x, \omega_y)$ is the Fourier transform of the spread function or impulse response $s(x, y)$. The magnitude of the OTF is called the modulation transfer function. It is a measure of how well the system accurately reproduces the scene. In other words, MTF is a measure of the ability of an imaging component (or system) to transfer the spatial modulation from the object to the image plane. Spatial modulation of light irradiance is related to image quality. If there is no such modulation, irradiance is uniform and there is no image. The highest spatial frequency that can be accurately reproduced is the system cutoff frequency. The maximum frequency an imager can detect without aliasing is defined as the Nyquist frequency, which is equal to one over two times the pixel pitch, i.e., $1/(2 \times \text{pixel pitch})$.

The phase transform function (PTF) is not nearly as important to resolution as MTF, but nevertheless it often cannot be neglected [11]. The spatial phase determines the position and orientation of the image rather than the detail size. If a target is displaced bodily in the image plane such that each part of the target image is displaced by the same amount, the target image is not distorted. However if portions of the image are displaced more or less then other portions, then the target image is distorted. This information is contained in the PTF.

The physical implications of the OTF are very analogous to those of the electrical transfer functions. Both permit determination of the output for any given input. In both cases, the Fourier transform of a delta function is a constant (unity) that contains all frequencies—temporal in electronics and spatial in optics—at constant (unity) amplitude. Therefore, a delta function input permits an indication of the frequency response of the system, that is, which frequencies pass through the system unattenuated and which frequencies are attenuated and by how much. A perfect imaging system (or any other transform system) requires resemblance between input and output; this would require infinite bandwidth so that an impulse function input would result in an impulse function output. Therefore, to obtain a point image for a point object, an infinite spatial frequency bandwidth is required of the imaging system. Physically, this means that because of the diffraction effects of optical elements such as camera lenses or pixels, these elements must be of infinite diameter to eliminate diffraction at edges. This, of course, is not practical. As a consequence, the smaller the aperture the greater the relative amount of light diffracted and the poorer the image quality.

The mathematical concept of the transfer function is accompanied by a real physical phenomenon called "contrast". For a one-dimensional sinusoidally varying object wave,

$$I_{OB}(x) = b + a \cos 2\pi f_{x0} x$$

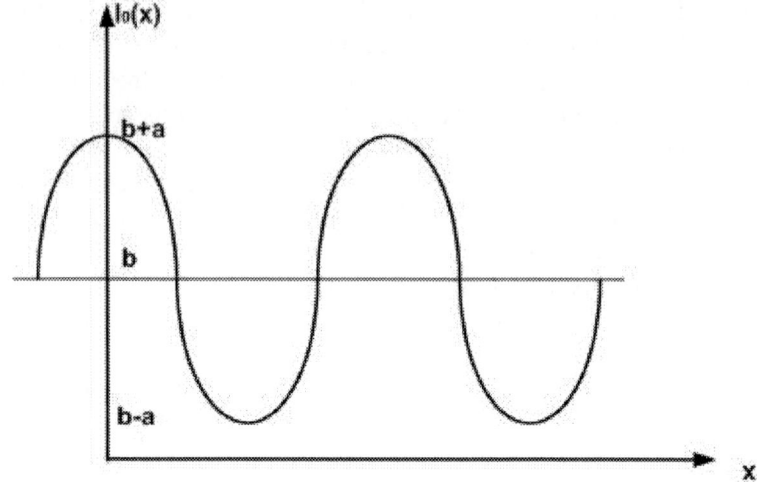

Figure 2-1. One-dimensional object irradiance.

The modulation contrast object (MCO), which describes the contrast in the object plane, is defined as

$$MCO = \frac{I_o\max - I_o\min}{I_o\max + I_o\min} = \frac{a}{b} \tag{2.2}$$

In the image plane,

$$I_{image}(x') = \int_{-\infty}^{\infty} I_{OB}(x)\, s(x - x')\, dx \tag{2.3}$$

Thus,

$$I_{IM}(x') = \mathbf{b}\,S(0)\left[1 + \frac{\mathbf{a}\,|S(f_{x0})|\cos\left[2\pi f_{x0}x' + \Phi(f_{x0})\right]}{\mathbf{b}\,S(0)}\right] \tag{2.4}$$

where $S(f_{x0})$ is the Fourier transform of the spread function, and $S(0)$ is the Fourier transform for the DC-level spread function. For a linear imaging system, the image of a cosine is thus a cosine of the same spatial frequency, with possible changes in the phase of the cosine $\Phi(f_{x0})$, which imply changes in position. The modulation contrast image (MCI), i.e., the contrast function in the image plane, is

$$MCI = \frac{I_{IM} \max - I_{IM} \min}{I_{IM} \max + I_{IM} \min} = \frac{\frac{a}{b}\left|S(f_{x0})\right|}{S(0)} \qquad (2.5)$$

The modulation contract function (MCF) or the contrast transfer function (CTF) is defined as the ratio of the image (output) modulation contrast to that of the object (input). For a cosine object,

$$MCF = CTF = \frac{\left|S(f_{x0})\right|}{S(0)} = MTF \qquad (2.6)$$

The MTF characteristics of a sensor thereby determine the upper limits of the image quality, i.e., the image resolution or sharpness in terms of contrast as a function of spatial frequency, normalized to unity at zero spatial frequency.

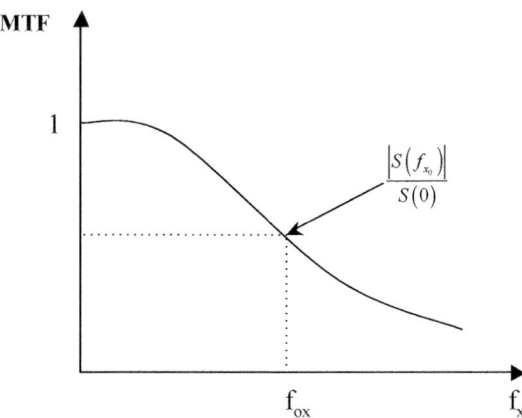

Figure 2-2. A typical MTF curve indicating which frequencies pass through the system unattenuated and which frequencies are attenuated and by how much.

2.1.2 Point spread function: the dependence on pixel size

The MTF varies in general as a function of illumination wavelength and can vary as a function of illumination intensity. Therefore, the conditions and techniques used for MTF measurement should be chosen based on the ultimate application of the focal plane array (FPA). A detailed analysis and comparison of the various MTF techniques was performed by T. Dutton, T. Lomheim et al. [12].

The point spread function (PSF, also known in the electro-optical community as the aperture response profile) is the spatial analogue of the two-dimensional MTF. A direct measurement of the PSF provides maximum insight into the vertical layer structure of the photosensor layout. It permits the experimental determination of the photodiode aperture response behavior (the geometric MTF) and the APS depletion/diffusion structure onset (the diffusion MTF).

To measure the PSF, it is important to determine the size at which an object can be considered a point object. The object-pixel is the smallest element recordable in the image space. The brightness value represents average irradiance over that small portion of the image scene. Object-pixel size is often related to the detector size. If only a portion of the detector is illuminated, the output current is equivalent to that obtained for the same total radiant power absorbed by the detector but averaged over the entire detector area. No detail smaller than that object-pixel can be resolved. Because of mechanical, atmospheric, and detector imperfections in the image system, the actual spread function obtained from the measured object-pixel is usually larger than the true object-pixel; thus, it represents an overall system response or spread function. Note that the object-pixel size strongly affects the system MTF. If an object-pixel represents a point image, then the pixel size and shape determines the minimum spread function. A best-case transform function for such an imaging system is thus a normalized Fourier transform of the pixel shape. For a rectangular pixel array, the MTF is determined then by the pixel (and pitch) dimensions by the well-known sinc formula:

$$H(w) = \frac{1}{p} \int_{-\frac{p}{2}}^{\frac{p}{2}} h(x) e^{jkx} dx = \frac{1}{p} \int_{-\frac{a}{2}}^{\frac{a}{2}} 1 \cdot e^{jkx} dx = \frac{a}{p} \frac{sin(\frac{ka}{2})}{(\frac{ka}{2})} \qquad (2.7)$$

where p is the pitch size, a is the sensor size, k is the angular frequency, and $h(x)$ is the impulse response. In this example, a maximum value occurs at $(2a)^{-1}$, when the sinc function reaches its first zero. As a decreases, the pixel size MTF broadens. A decrease in pixel size means that smaller details can be resolved, corresponding to a larger frequency bandwidth. The smaller the pixel dimensions, the larger the spatial-frequency bandwidth. A good tool for measuring the MTF of the detector or of the system (taking into consideration the imperfections of the light propagation channel) is a point source corresponding to the image plane size.

2.1.3   CMOS APS MTF modeling: a preview

Solid-state imagers are based upon rectangular arrays of light-sensitive imaging sites, also called picture elements or pixels. In CMOS APS arrays, the pixel area is constructed of two functional parts. The first part, that has a certain geometrical shape, is the sensing element itself: the active area that absorbs the illumination energy within it and turns that energy into charge carriers. Active pixel sensors usually consist of photodiode or photogate arrays [13–16] in a silicon substrate. Each imaging site has a depletion region of several micrometers near the silicon surface. Perfect collection efficiency is assumed for carriers at or within the depletion region, and therefore any photocarrier generated in this depletion region is collected at the imaging site. The second part of the pixel area is the control circuitry required for readout of the collected charge. The fill factor (FF) for APS pixels is less than 100 percent, in contrast to CCDs where the FF can approach 100%. The preferred shape of the active area of a pixel is square. However, designing the active area as a square can reduce the fill factor. Since it influences the signal and the SNR, the fill factor should be as high as possible. *Figure 2-3* shows a pixel with an L-shaped active area, which is the type most commonly used.

Photon absorption in the silicon depends on the absorption coefficient α, which is a function of the wavelength. Blue light (with wavelengths of $\lambda \approx 0.4$ μm) is strongly absorbed in the first few micrometers of silicon, since α is large in this spectral region. Longer wavelengths (such as $\lambda \approx 0.6$ μm) have a smaller absorption coefficient, which means more of the photocarriers can be generated outside the depletion regions. Before they are lost to a bulk recombination process, these carriers can diffuse to the original imaging site or to a nearby site where they are collected. However, the imagers lose resolution as the result of this diffusion process. *Figure 2-4* shows a schematic cross-section of several imager sites, indicating the depletion-region boundaries.

Theoretical calculations to model the effects of the photogenerated minority carriers' spatial quantization, transfer efficiency and crosstalk in CCDs and CISs have been described over the years, [1–9]. It has always been assumed that the theoretical model includes the solution for the continuity equation

$$DV^2n + \frac{n}{\tau} = Nf \cdot \alpha \exp[-\alpha z] \qquad (2.8)$$

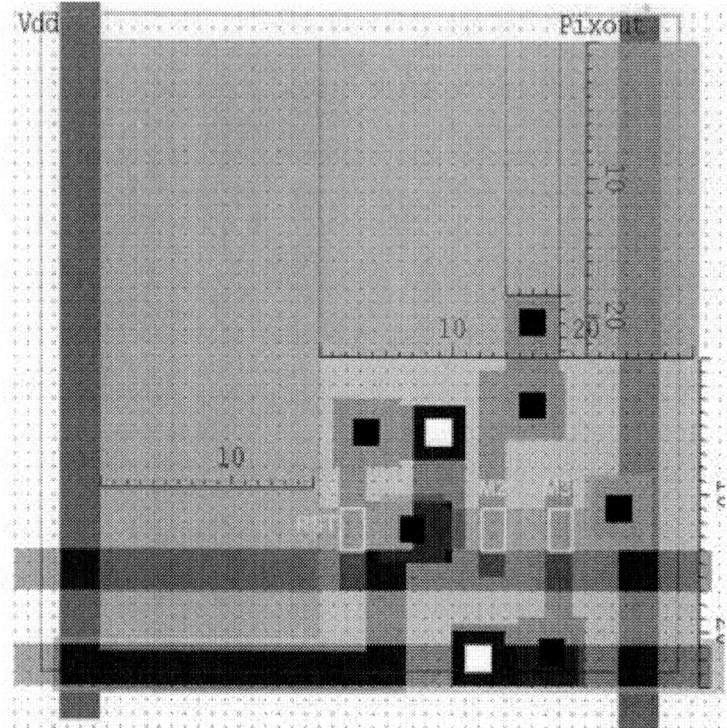

Figure 2-3. Example layout of a pixel design with an L-shaped active area.

where *n* represents the minority carriers concentration, *Nf* is the flux transmitted to the substrate of the sensor at some position *x*, *α* is the absorption coefficient, *z* the depth within the semiconductor, *τ* the minority carrier lifetime and *D* the diffusion coefficient.

The standard diffusion MTF models must be modified to account for modern layer structure as discussed by Blouke and Robinson [3] and Stevens and Lavine [7]. The latter researchers developed an approach wherein the pixel aperture and diffusion components of the MTF are treated in a unified manner. It was shown that the multiplication of the diffusion and aperture MTF for the purpose of arriving at the overall sensor MTF is only valid for the special case in which the pixel aperture is equal to its size or pitch, i.e., for a 100% fill factor. In other words, the unification of the geometrical and diffusion effects is necessary for the correct MTF representation, i.e., $MTF_u = MTF_g * MTF_{diff}$. Lin, Mathur and Chang [17] show that the CCD-based diffusion models do not work well for APS imagers, since the latter devices have field-free regions between and surrounding the pixel photodiodes that contribute to diffusion currents.

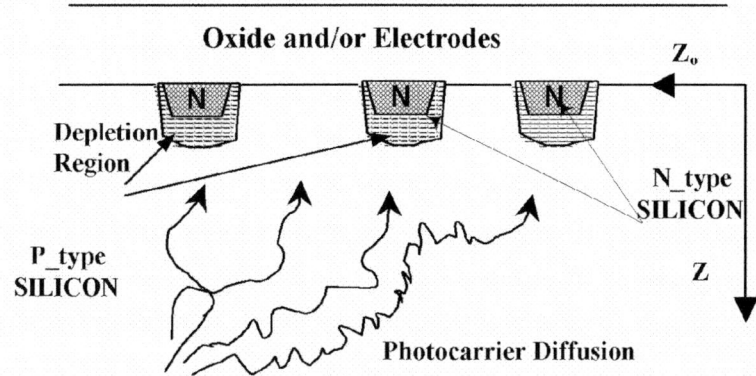

Figure 2-4. A schematic cross-section of several imager sites, illustrating the dependence of the diffusion distance on the photodiode geometry.

This chapter is a logical extension of a theoretical analysis recently presented by Yadid-Pecht [10], in which the theoretical MTF for the active-area shape of a general pixel was calculated and compared with experimental data. It was shown that the active-area shape contributes significantly to the behavior of the overall MTF of a CMOS imager.

The general expression for the MTF for the connected shape pixel is

$$MTF(k) = \sum_{i=1}^{m} \left| e^{jk\frac{a_1}{2}} \left[\frac{Ai}{P} \cdot (a_i - a_{i-1}) \cdot n_i \cdot e^{jk\frac{a_i}{2}} \cdot sinc(\frac{k}{2}(a_i - a_{i-1})) \right] \right|$$

(2.9)

where

$$n_i = \frac{a_i}{\sum_{j=1}^{} A_j \cdot (a_j - a_{j-1})} \quad \text{and} \quad a_0 = 0$$

(See *Figure* 2-5, left side.)

In the analysis for L-shaped active areas [10], four parameters are inserted: a_1, a_2, A_1 and A_2. In general, A_i is the modulation amplitude and a_i is the active area length. For a 2-D array, a multiplication of the MTF in both directions is required. To calculate the MTF in the y direction, a simple variable change is performed. The modulation amplitude in the x direction is

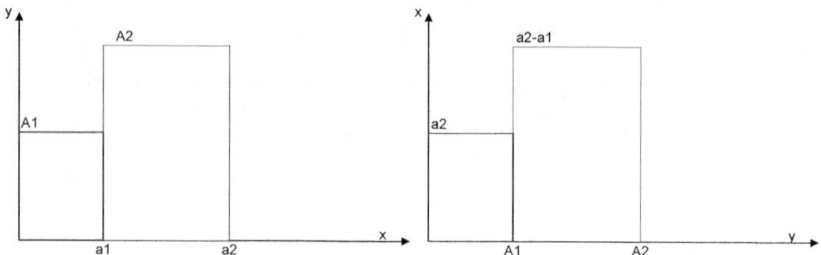

Figure 2-5. (Left) Schematic description of the L-shaped detector. *(Right)* Schematic description of the L-shaped parameters in the *y* direction (after [10]).

actually the length in the *y* direction, while the length in the *x* direction is the amplitude modulation in the *y* direction (see *Figure 2-5*, right side).

In that work, analysis of the MTF for the general pixel active area shape was considered. An analytical solution was derived for the most commonly used shape in practical pixel designs, the L-shaped detector. The actual PSF was obtained experimentally via sub-pixel scanning, and the MTF was calculated accordingly from the measurements for the different pixel designs.

A good general agreement was found between this calculated MTF and the theoretical expectations. The differences that remained between the actual and simulated MTFs were mostly due to other factors in the design and process, such as diffusion and crosstalk (see *Figure 2-6*).

Here we present a more comprehensive model, which takes into account the effect of the minority-carrier diffusion together with the effect of the pixel active-area shape on the overall CMOS-APS MTF. This is especially important for APS design, where the fill factor is always less than 100%.

2.2   Experimental details

Our model is based on the measurements of responsivity variation on a subpixel scale for the various APS designs. These measurements were reported in [10] and will be described here briefly. An optical spot approximately 0.5 µm in diameter (He–Ne laser, $\lambda = 0.6$ µm) was used to scan the APS over a single pixel and its immediate neighbors in a raster fashion. This work dealt only with the ideal situation, where light was projected directly onto the sensor pixel. Therefore, effects on the MTF due to the presence of an optical stack of oxides such as light piping [18] were not considered. In addition, the spot size of the laser was small compared to the pixel size; therefore, the effects of laser spot profile on the measured PSF were not considered.

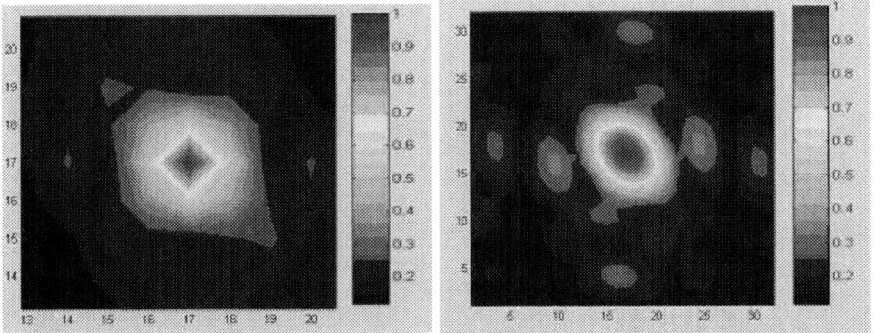

Figure 2-6. (Left) MTF obtained from the scanned PSF. *(Right)* MTF obtained with the analytical geometrical model.

The data acquisition was taken at the center point; i.e., the central pixel was read out at each point of the scan (see *Figure 2-7*). The signal obtained as a function of the spot position provided a map of the pixel response.

Only the central pixel is assumed to be active. The "window region", i.e., the photodiode (or photogate) of the active pixel is the only region where the wide (non-zero bias) depletion layer exists and the photocarrier collection occurs. The incoming laser light generates (at a certain depth according to the exponential absorption law) electron-hole pairs, i.e., minority charge carriers. Diffusion of these charge carriers occurs with equal probability in all directions, with some diffusing directly to the depletion region where they subsequently contribute to the signal. Therefore, the PSF obtained in the window region (see *Figure 2-8*) is due to the detection of those charge carriers that successfully diffused to the depletion region. The value at each point represents the electrical outcome of the three-dimensional photocarrier diffusion (i.e., the integration over the depth at each point) from that point to the depletion. Thus, the 2-D signal map plane obtained in the experiment can be generally considered as a "diffusion map" of the 3-D diffusion in the device.

2.3   Physical analysis

The PSF data obtained from the actual pixel measurements for a square, a rectangular, and an L-shaped active area were examined using scientific analysis software. Note that without the limitation of generality, this chapter demonstrates only the results for an L-shaped active area pixel as an

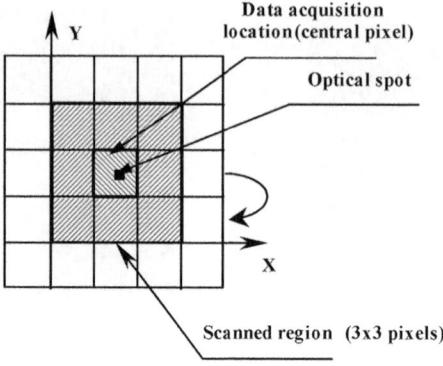

Figure 2-7. Geometry of the generalized experiment. The squares represent the APS subarray. The optical spot (the small black square) was scanned over the array in a raster fashion within a specified region (the shaded squares).

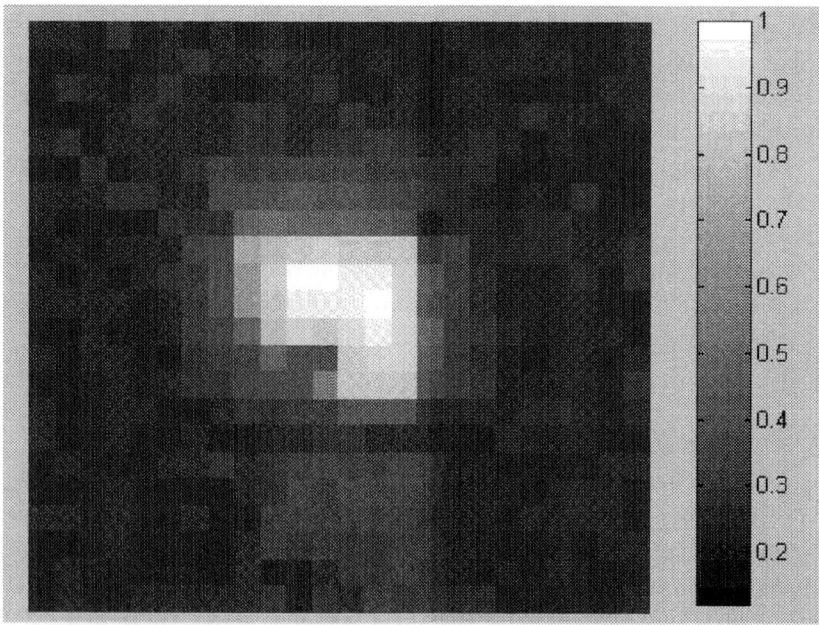

Figure 2-8. Plot of the actual measured PSF for the L-shaped pixel design (after [10]). The lightest areas indicate the strongest response.

example. In the following well-known steady-state one-dimensional transport equation of excess minority carriers in the semiconductor,

$$\frac{\partial n_p}{\partial t} = -\frac{n_p - n_{p0}}{\tau_n} + D_n \frac{\partial^2 n_p}{\partial x^2} \tag{2.10}$$

the electron recombination rate can be approximated by $(n_p - n_{p0})/\tau_n$. Here n_p is the minority carrier density, n_{p0} is the thermal equilibrium minority carrier density, τ_n is the electron (minority) lifetime, and D_n is the electron diffusion coefficient. The solution is given by:

$$n_p(x,t) = \frac{N}{\sqrt{4\pi D_n t}} \exp\left[-\frac{x^2}{4D_n t} - \frac{t}{\tau_n}\right] + n_{p0} \tag{2.11}$$

where N is the number of electrons generated per unit area.

Based on this solution, the PSF data acquired from each actual scanning was fitted. For each scanned pixel a set of fittings was performed. *Figure 2-9* indicates examples of planes on which fitting was performed for the L-shaped pixel. A'—A" is one of the cross-sections on which a fit was performed.

All the pixels are situated on a common substrate. The photocarrier diffusion behavior within the substrate is therefore the same across a given pixel array. From the generalization of the fitting results for all pixels, the common functional dependence is derived with common parameters that describe the diffusion process in the array. The two-dimensional approach described earlier allows the desired physical parameters that correspond to the actual three-dimensional diffusion to be obtained.

The model used for fitting is

$$y = y_0 + \frac{A}{W\sqrt{\frac{\pi}{2}}} \exp\left[-\frac{2(x - x_c)^2}{W^2} - 1\right] \tag{2.12}$$

where x_c is the center, W the width, and A the area. The term "–1" describes the case of $\tau = t$. It is generally possible to perform fittings with values of t equal to various fractions of τ, so that the best correspondence is obtained for each window shape. The relevant parameters were derived from a comparison of equations (2.11) and (2.12):

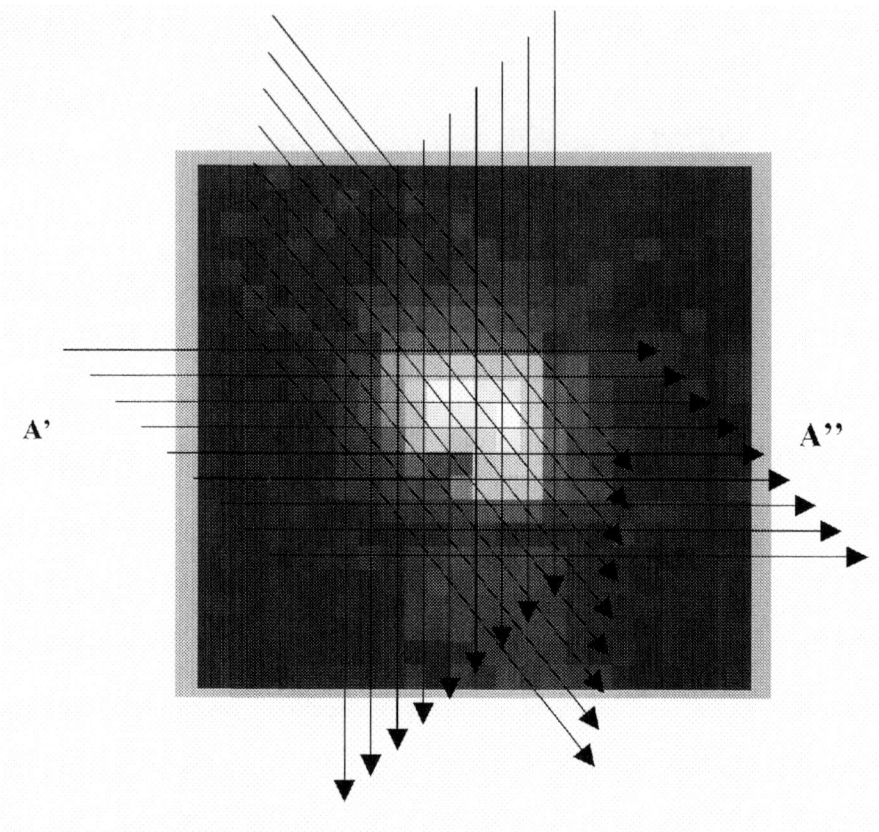

Figure 2-9. Plot of the actual measured PSF for the L-shaped pixel design (after [10]). Cross-sections used for fitting are located along the arrows, normal to the layout surface. The lightest areas indicate the strongest response, as in *Figure 2-8*.

$$W = 2\sqrt{2D_n\tau}, \;\Rightarrow\; L_{eff} = \sqrt{D_n\tau} \qquad (2.13)$$

where L_{eff} is the characteristic effective diffusion length. The common average result obtained with this approach was $L_{eff} \approx 24$ μm.

Figure 2-10 shows the curve-fitting function and the actual data points for an example cross-section.

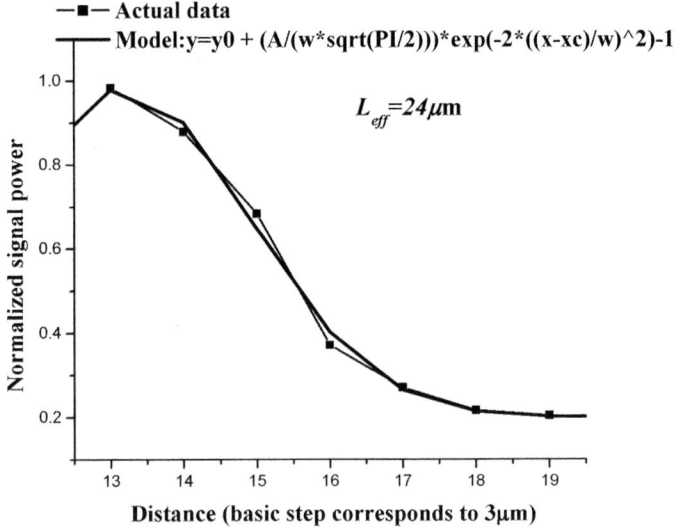

Figure 2-10. Functional analysis of the measured point spread function. The example data correspond to the A'—A'' cross-section in *Figure 2-9.*

2.4 The unified model description

Based on the analysis described for the experimental data and the actual layout of the pixel array, a unified numerical model was constructed that included both the effect of photocarrier diffusion within the substrate and the effect of the pixel sampling aperture shape and size within the pixel array. This model produces the PSF of the pixel empirically. The extracted parameters are used for the creation of a 2-D symmetrical kernel matrix (since there is no diffusion direction priority within the uniform silicon substrate). The convolution of this matrix with the matrix representing the pure geometrical active area shape artificially produces the response distribution in the spatial domain. Note that the spread function obtained in this way creates a unified PSF; i.e., this method enables modeling of the pixel spatial response, which can subsequently be compared with the genuine PSF obtained by real measurements.

The dimensions of the kernel matrix are important. *Figure 2-11* explains the rationale for choosing these dimensions. At the points corresponding to the kernel dimension (i.e., points 7 to 9), a minimum is reached for both the

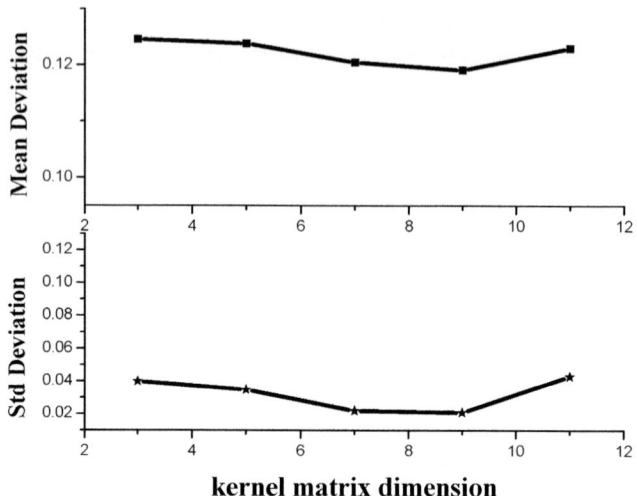

Figure 2-11. Dependence of the mean and standard deviations on the kernel matrix dimensions.

mean and standard deviation functions obtained from the comparison between the modeled PSF and the scanned PSF.

In both cases, these kernel matrix dimensions equal the physical pixel size used in this representation. Thus, we conclude that the diffusion occurs primarily within the pixel, i.e., $L_{eff} \approx 24.4$ μm. The value for this parameter that was here obtained directly from the model is the same as the one previously obtained analytically by fitting the data.

The experimental results have been compared with predictions for several practical designs. A fair correspondence of the simulated PSF with the measured PSF was generally obtained. A thorough discussion of the results follows in the next section.

2.5   Results and discussion

To compare the PSF and the MTF of practical pixels with square, rectangular and L-shaped active areas, the two-dimensional MTF of these cases were calculated, simulated, and compared with the measurements. The measurements currently used for the analysis were obtained with relatively old CMOS 1.2 μm process chips as described in [10]. However, the analysis presented here is general in nature, and so a similar MTF behavioral trend is

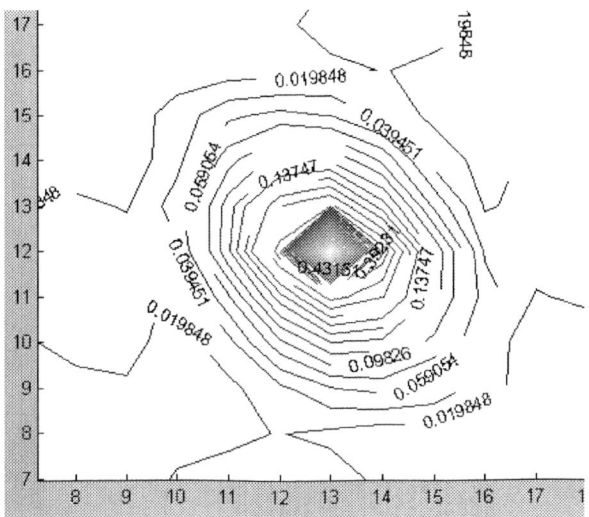

Figure 2-12. The MTF contour plot calculated from the PSF obtained by laser scanning of the L-shaped pixel design (after [10]).

expected for scalable CMOS processes. The design consisted of an APS sensor with differently shaped pixel active areas: a square-shaped active area with a fill factor of about 8%, a rectangular-shaped active area with a fill factor of 31%, and an L-shaped design with a fill factor of around 55%.

An example of the L-shaped layout is shown in Figure *2-3*. Figure *2-8* shows the corresponding point spread function map obtained by laser scanning in subpixel resolution (after [10]). Figure *2-12* shows the corresponding MTF, calculated via a 2-D Fourier transform.

Figure *2-13* represents the PSF map obtained from the unified computer model and Figure *2-14* shows the corresponding MTF contour plot.

Figure *2-15* and Figure *2-16* show the comparisons of the model and actual PSF plots. Figure *2-15* gives the difference between the measured PSF and the pure geometrical PSF (i.e., the PSF resulting when the active-area pixel response is represented by unity in the active area and zero otherwise [10]) for a specific pixel. The maximum difference observed is about 20% of the maximum pixel response. Figure *2-16* compares the measured PSF and the PSF obtained from the unified model, which takes into account the diffusion effects. The maximum difference observed is about 3% of the maximum pixel response.

Table 2-1 compares the extracted diffusion length values and confirms the universality of the method described in this chapter. The calculations are based on data extracted from relevant literature sources. The same model can be used for any process design.

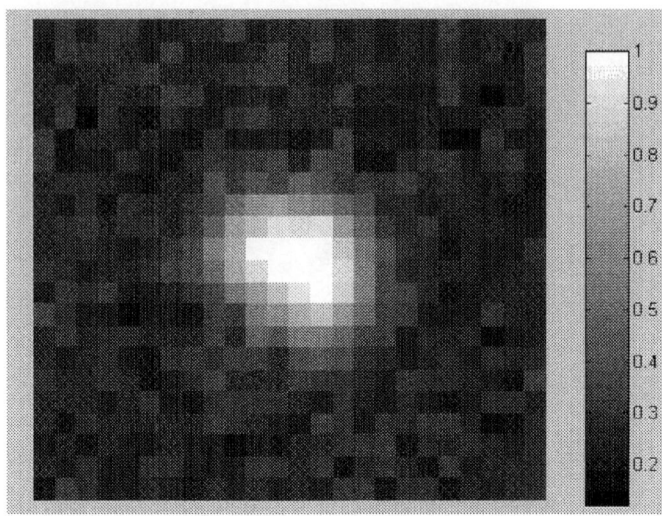

Figure 2-13. Plot of the PSF obtained from the unified model for the L-shaped pixel design. The lightest areas indicate the strongest response.

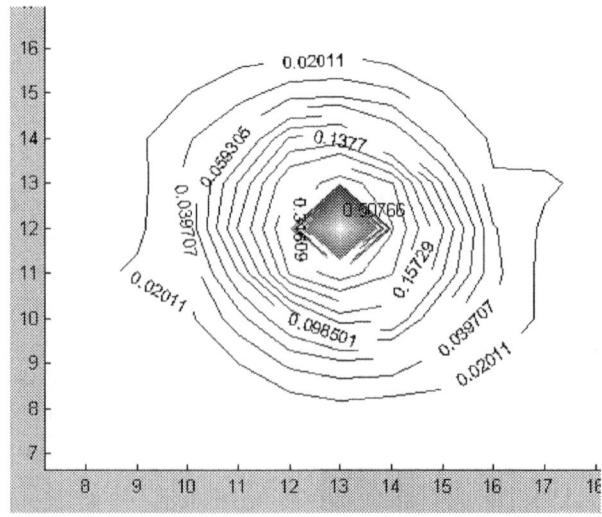

Figure 2-14. Contour plot of the MTF simulation result for the L-shaped pixel design.

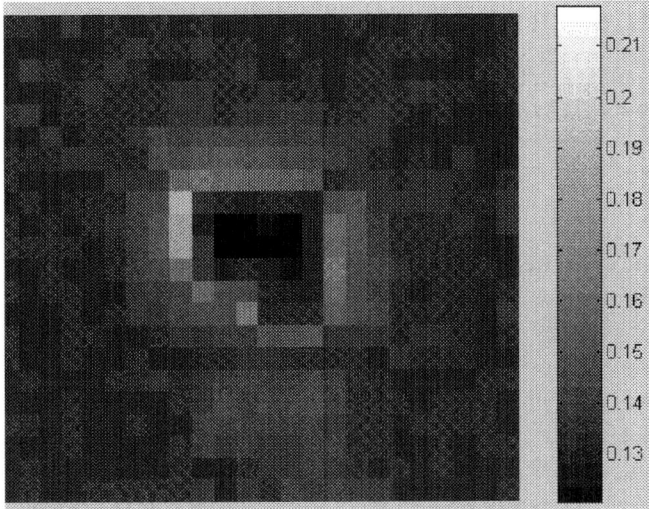

Figure 2-15. The difference between the pure geometrical PSF and the PSF obtained by scanning the L-shaped pixel design. The lightest areas indicate the strongest response. The maximum difference is about 20% of the maximum pixel response.

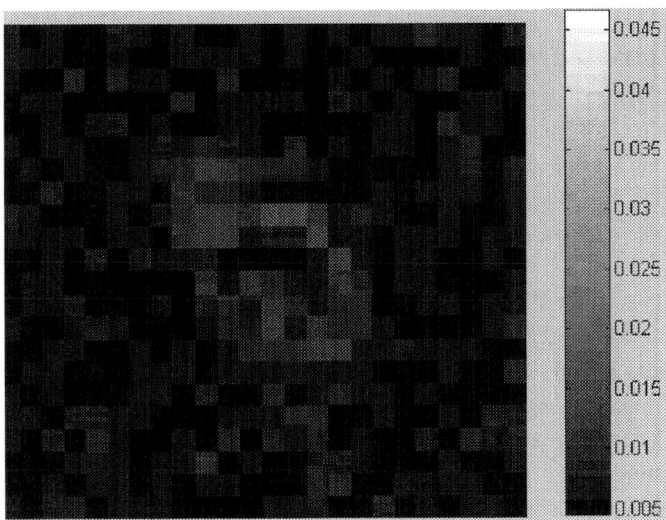

Figure 2-16. The difference between the unified model PSF and the PSF obtained by scanning the L-shaped pixel design. The lightest areas indicate the strongest response. The maximum difference is about 3% of the maximum pixel response.

Table 2-1. Comparison of extracted diffusion length values.

Diffusion length	Value obtained (μm)
Extracted by function fitting	~24
Extracted by kernel optimization	~24.4
Calculated $L_{eff} = \sqrt{D_n \tau}$ ($D_n = 37.5$ cm^2/sec (300K) [13]; $\tau = 20$ psec [14])	~27

It has already been shown [10] that the active-area shape contributes significantly to the behavior of the overall MTF. However, there are essential differences between the actual and geometrical MTFs. The unified model presented here gives better agreement between the modeled and the actually measured PSF and MTF. Some difference can still be seen (*Figure 2-16*) between the actually scanned and the modeled PSF matrices. However, these differences are confined to the background level, and on average are calculated to be less than 1%. They occur due to other factors in the design and process, such as optical crosstalk [18]. Optical crosstalk results from interference in the oxide level, especially between metal lines, and has an effect on the overall MTF [1, 4, 6]. This factor would have a larger effect as the pixel size scales in multi-level metal processes [19].

In addition, the effects of the laser spot profile (used for scanning) on the resulting PSF should be considered for smaller pixels.

2.6   Summary

Based on the analysis of subpixel scanning sensitivity maps, a unified model for estimating the MTF of a CMOS-APS solid-state image sensor was developed. This model includes the effect of photocarrier diffusion within the substrate in addition to the effects of in the pixel sampling aperture shape and size.

Minority-carrier diffusion length, which is characteristic for the process, was extracted for various active-area pixels via several different methods.

The comparison of the simulation results with the MTF calculated from the PSF direct measurements of actual pixels confirmed that the two determining factors that affect the overall MTF behavior were the active-area shape and the minority-carrier diffusion effect.

The results also indicated that a reliable estimate of the degradation of image performance is possible for any pixel active-area shape; therefore, the tradeoffs between conflicting requirements, such as signal-to-noise ratio and MTF, can be compared for each pixel design and better overall sensor performance can ultimately be achieved.

The unified model enables a design-enabling optimal-pixel operation (in the MTF sense) based on readily available process and design data. Thus, the model can be used as a predictive tool for design optimization in each potential application.

The proposed model is general in nature. However, evolving technologies will cause stronger scaling effects, which will necessitate further model enhancements. In addition, the unique submicron scanning system [20, 21] will allow exploration of the strong wavelength dependence of the diffusion component of the MTF, and it is hoped that aperture and lateral diffusion effects can be separated via empirical measurements.

BIBLIOGRAPHY

[1] D. H. Seib, "Carrier diffusion degradation of modulation transfer function in charge coupled imagers," *IEEE Trans. Electron Devices*, vol. 21, ED-3, pp. 210–217, 1974.

[2] S. G. Chamberlain and D. H. Harper, "MTF simulation including transmittance effects of CCD," *IEEE Trans. Electron Devices*, vol. 25, ED-2, pp. 145–154, 1978.

[3] M. Blouke and D. Robinson, "A method for improving the spatial resolution of frontside-illuminated CCDs," *IEEE Trans. Electron Devices*, vol. 28, pp. 251–256, Mar. 1981.

[4] J. P. Lavine, E. A. Trabka, B. C. Burkey, T. J. Tredwell, E. T. Nelson and C. N. Anagnosyopoulos, "Steady-state photocarrier collection in silicon imaging devices," *IEEE Trans. Electron Devices*, vol. 30, ED-9, pp. 1123–1134, Sept. 1983.

[5] J. P. Lavine, W. Chang, C. N. Anagnosyopoulos, B. C. Burkey and E. T. Nelson, "Monte Carlo simulation of the photoelectron crosstalk in silicon imaging devices," *IEEE Trans. Electron Devices*, vol. 32, ED-10, pp. 2087–2091, 1985.

[6] E. G. Stevens, "A unified model of carrier diffusion and sampling aperture effects on MTF in solid-state image sensors," *IEEE Trans. Electron Devices*, vol. 39, ED-11, pp. 2621–2623, 1992.

[7] E. G. Stevens and J. P. Lavine, "An analytical, aperture and two-layer diffusion MTF and quantum efficiency model for solid-state image sensors," *IEEE Trans. Electron Devices*, vol. 41, ED-10, pp. 1753–1760, 1994.

[8] D. Kavaldjiev and Z. Ninkov, "Subpixel Sensitivity Map for a Charge Coupled Device sensor," *Opt. Eng.*, vol. 37, no. 3, pp. 948–954, Mar. 1998.

[9] T. O. Körner and R. Gull, "Combined optical/electric simulation of CCD cell structures by means of the finite-difference time-domain method," *IEEE Trans. Electron Devices*, vol. 47, ED-5, pp. 931–938, May 2000.

[10] O. Yadid-Pecht, "The geometrical modulation transfer function (MTF) for different pixel active area shapes," *Opt. Eng.*, vol. 39, no. 4, pp. 859–865, 2000.

[11] N. S. Kopeika, *A System Engineering Approach to Imaging*, Bellingham, WA, USA: SPIE Press, 1998.

[12] T. Dutton, T. Lomheim and M. Nelsen "Survey and comparison of focal plane MTF measurement techniques," *Proc. SPIE*, vol. 4486, pp. 219–246, 2002.

[13] O. Yadid-Pecht, "CMOS Imagers" (course notes), Ben Gurion University, Beer-Sheva, Israel, 2000.

[14] S. M. Sze, *Physics of Semiconductor Devices*, New York: J. Wiley and Sons, 1981.

[15] P. Bhattacharya, *Semiconductor Optoelectronic Devices*, Upper Saddle River, NJ: Prentice Hall, 1993.

[16] T. Spirig, "Smart CCD/CMOS based image sensor with programmable real time temporal, and spatial convolution capabilities for application in machine vision and optical metrology," Dissertation #11993, ETH, Switzerland.

[17] C. Lin et al., "Analytical charge collection and MTF model for photodiode-based CMOS imagers," *IEEE Trans. Electron Devices*, vol. 49, ED-5, pp. 754–761, May 2002.

[18] D. Ramey and J. T. Boyd, "Computer simulation of optical crosstalk in linear imaging arrays," *IEEE J. Quantum Electron.*, vol. 17, pp. 553–556, Apr. 1981.

[19] H. Wong, "Technology and device scaling considerations for CMOS imagers," *IEEE Trans. Electron Devices*, vol. 43, no. 12, pp. 2131–2142, Dec. 1996.

[20] I. Shcherback, B. Belotserkovsky, A. Belenky and O. Yadid-Pecht, "A unique sub-micron scanning system use for CMOS APS crosstalk characterization," in *SPIE/IS&T Symp. Electronic Imaging: Science and Technology*, Santa Clara, CA, USA, Jan. 20–24, 2003.

[21] I. Shcherback and O. Yadid-Pecht, "CMOS APS crosstalk characterization via a unique sub-micron scanning system," *IEEE Trans. Electron Devices*, vol. 50, no. 9, pp. 1994–1997, Sept. 2003.

Chapter 3

PHOTORESPONSE ANALYSIS AND PIXEL SHAPE OPTIMIZATION FOR CMOS APS

Igor Shcherback and Orly Yadid-Pecht
The VLSI Systems Center
Ben-Gurion University
P.O.B. 653 Beer-Sheva 84105, ISRAEL

Abstract: A semi-analytical model has been developed for the estimation of the photoresponse of a photodiode-based CMOS active pixel sensor (APS). This model, based on a thorough analysis of experimental data, incorporates the effects of substrate diffusion as well as geometrical shape and size of the photodiode active area. It describes the dependence of pixel response on integration photocarriers and on conversion gain. The model also demonstrates that the tradeoff between these two conflicting factors gives rise to an optimum geometry, enabling the extraction of a maximum photoresponse. The dependence of the parameters on the process and design data is discussed, and the degree of accuracy for the photoresponse modeling is assessed.

Key words: CMOS image sensor, active pixel sensor (APS), diffusion process, quantum efficiency, parameter estimation, optimization, modeling.

3.1   Introduction

This work is a logical continuation of the pixel response analysis published by Yadid-Pecht et al. [1]. In this chapter, a pixel photoresponse is analyzed and quantified to provide the information necessary for its optimization. A novel way for the determination and prediction of the imager quantum efficiency (QE) is also presented. In this method, the QE is broadly interpreted to be dependent on the process and design data, i.e., on the pixel geometrical shape and fill factor. It is worth noting that QE, which is one of the main figures of merit for imagers, has been considered a whole pixel characteristic without any specific attention to the internal pixel geometry. However, it is useful to divide it into the main and diffusion parts. Even though the active area has the most effect on the output, the substrate parts could account for up to 50% of the total output signal. The derived expression exhibits excellent agreement with the actual measurements obtained from a 256 × 256 CMOS active pixel sensor (APS) imager. The simplicity and the accuracy of the model make it a suitable candidate for implementation in photoresponse simulation of CMOS photodiode arrays.

This chapter presents the proposed photoresponse model and shows the correspondence between the theory and the experimental data. The model is then used to predict a design that will enable maximum response in different scalable CMOS technologies.

3.1.1   General remarks on photoresponse

Single-chip electronic cameras can be fabricated using a standard CMOS process; these cameras have the advantages of high levels of integration, low cost, and low power consumption [2–13]. However, getting an acceptable response from the currently available CMOS structures is a major difficulty with this technology.

A particular problem encountered by designers of vision chips in standard CMOS technologies is that foundries do not deliver characterization data for the available photosensors. Thus, designers are forced to use simplistic behavioral models based on the idealized descriptions of CMOS-compatible photosensors descriptions. This may be enough for chips intended to process binary images. However, the characteristics of real devices largely differ from their idealized descriptions, being, on the other hand, strongly dependent on the fabrication process. Consequently, using these idealized models yields very inaccurate results whenever the analog content of the incoming light (regarding both intensity and wavelength) is significant for signal processing; i.e., chips whose behavior is anticipated on the basis of such simplified models most likely will not accomplish the specifications. In this scenario, designers of vision chips are confronted by the necessity to fully characterize the CMOS-compatible photosensors by themselves. This is not a simple undertaking and requires expertise that is not common among chip designers. This chapter is intended to help designers acquire the skills to accomplish this task.

In order to predict the performance of an image sensor, a detailed understanding is required of the photocurrent collection mechanism in the photodiodes that comprise the array.

The response of the detectors can be affected by different parameters that are dependent on the process or on the layout. Process-dependent parameters (such as junction depths and doping concentration) cannot be modified by the designer, whereas layout-dependent parameters (detector size and shape) are under the designer's control. Both types of parameters have to be considered when designing a test chip for characterizing visible-light detectors. All possible devices or structures available in the technology to detect visible light have to be included in the chip so that their actual behavior can be measured. In addition, these detectors should each be tested in different sizes and shapes to account for the layout-dependent parameters. Each detector should ideally be tested in as many sizes as possible, with a range of different perimeters (shapes) for each size. Furthermore, the chip should include as many copies of each detector configuration as possible to obtain statistical information on their response. Since detectors must be connected to bonding pads to have direct access to the detected signal, the cost of fabrication of such a chip would be very high.

A more realistic approach would use a limited number of discrete detectors with different sizes and shapes, but still covering all the possible device structures; several arrays of these detectors would be tested to provide the statistical information. Accordingly, the experimental data in this chapter were acquired from several APS chips fabricated in standard, scalable CMOS technology processes (e.g., standard 0.5 μm and 0.35 μm CMOS processes). Various topologies of the photosensitive area were implemented. All the pixels had a common traditional three-transistor type of readout circuitry (see *Figure 3-1*), enabling behavior identification for the different pixel types. Deviations in the device geometry were demonstrated to affect overall performance and thus these dependencies can be used as a predictive tool for design optimization.

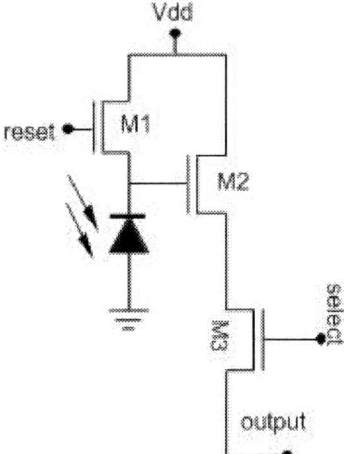

Figure 3-1. Traditional pixel readout circuitry. Transistor M2 is used as a source follower and the M2 gate is used to read out the photodiode voltage.

When light is incident on a diode (APS systems usually consist of photodiode or photogate arrays on a silicon substrate), photons absorbed in the depletion region of the p-n junction create electron-hole pairs that are separated by the high internal electric field. These new charges can be detected as an increase in the reverse current of the device or as a change in the voltage across it. Carriers generated in the bulk of the silicon, if they are less than a minority-carrier diffusion length away from the depletion region, can also contribute to the detected signal and thus increase the sensitive volume of the detector. The collection of photocarriers along the lateral edge of the photodiode is known as the peripheral photoresponse or the lateral photocurrent [14–17]. The overall signal generated by a pixel is therefore proportional to its geometry; i.e., the signal grows as photodiode dimensions increase. The total signal can be represented as the sum of the main (photodiode) response and the periphery response (due to the successfully collected diffusion carriers).

Figure 3-2 schematically shows cross-section of imager sites and indicates their depletion-region boundaries.

The response of these detectors is also affected by other parameters, such as technology scaling and detector size and shape. It is important to characterize all possible detectors for a range of sizes and shapes (independent of the circuitry) to obtain the photogenerated current levels for each case. All possible detectors also have to be characterized for each technology in order to find the optimum device and match the requirements of the signal processing circuitry.

In the case of CMOS APS, the charge-to-voltage conversion gain is typically dominated by the photodiode junction capacitance, which is composed of the bottom and sidewall capacitances. This capacitance can be expressed as

$$C_{jdep} = \frac{C_{J0B}A_D}{\left[1-\left(\frac{V_d}{\varphi_B}\right)\right]^{m_j}} + \frac{C_{J0sw}P_D}{\left[1-\left(\frac{V_d}{\varphi_{Bsw}}\right)\right]^{m_{jsw}}} \qquad (3.1)$$

where
- C_{jdep} represents the depletion capacitance of the p-n junction;
- C_{J0B} and C_{J0sw} represents zero-bias capacitances of the bottom and the sidewall components, respectively;
- V_d is the voltage applied to the photodiode;
- φ_B and φ_{Bsw} stand for the build-in potential of the bottom and the sidewalls respectively;

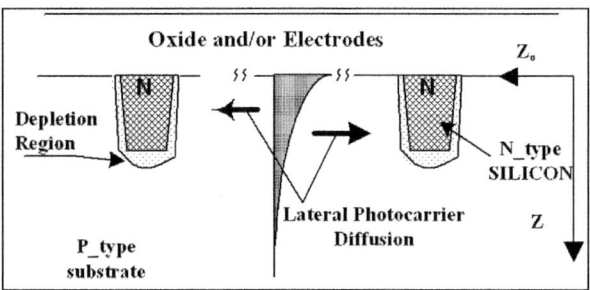

Figure 3-2. Cross-sections of imager sites and their depletion-region boundaries.

- m_j and m_{jsw} stand for the grading coefficients of the bottom and the sidewalls respectively;
- A_D represents the photodiode area (bottom component); and
- P_D represents the photodiode perimeter.

The conversion factor, $C_{gain} = q_{el}/C_{jdep}$ (in $\mu V/e^-$), is inversely proportional to the pixel geometry. The factor q_{el} is the electron charge.

The basic transfer characteristics of a photodetector are usually described by its quantum efficiency, which depends on the wavelength and describes the number of electrons generated and collected per incident photon. The quantum efficiency is related to the spectral response (in A/W) according to the equation:

$$SR(\lambda) = \frac{\lambda q}{hc}(\lambda) \tag{3.2}$$

If a photodiode is exposed to a spectral power density $\Phi(\lambda)$, the collected photocharge can be expressed as

$$Q = A_{eff} t_{int} \int SR(\lambda)\Phi(\lambda)d\lambda \tag{3.3}$$

with A_{eff} denoting the effective photoactive area of a pixel and t_{int} the integration time. Illumination is assumed to be constant during the exposure time.

The voltage swing that is obtained from the collected photocharge is inversely proportional to the integration capacitance, C_{int} (which equals the depletion capacitance C_{jdep} in CMOS APS), as follows:

$$V = \frac{Q}{C_{int}} = \frac{A_{eff} t_{int} q}{C_{int}} \int \frac{\lambda}{hc}\eta(\lambda)\,\Phi(\lambda)d\lambda \tag{3.4}$$

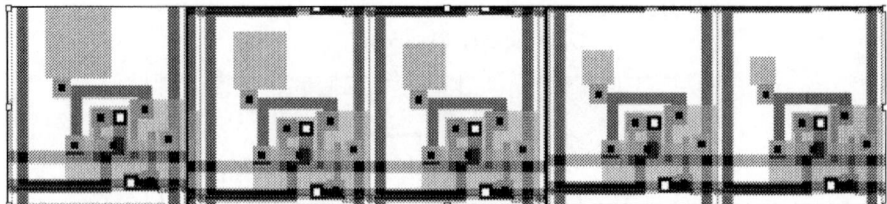

Figure 3-3. A sample from an otherwise identical square-shaped pixel set with decreasing active area (photodiode) dimensions. The photodiode areas in the full pixel set vary from 40 μm^2 down to 5.5 μm^2 and their perimeters from 23 μm down to 9.3 μm.

Consequently, pixel signal output is proportional to the product of the integrated photocarriers and the conversion gain. The tradeoff between these two conflicting parameters is most important. Indeed, it has already been shown that a lower active area contributes to a higher output signal. This is mainly due to an increase in conversion gain, but the active-area surroundings probably have an affect as well [1, 5, 10, 18]. However, a lower active area reduces the fill factor, which directly influences the signal and the signal-to-noise ratio. Both the active area and the fill factor should therefore be as large as possible.

3.2   Photoresponse model

Figure 3-3 shows a sample from a set of pixels of identical square shape but decreasing active area (photodiode) dimensions. The photodiode areas vary from 40 μm^2 down to 5.5 μm^2, and their perimeters vary from 23 μm down to 9.3 μm.

Figure 3-4 displays the signal output obtained for illumination at three different wavelengths with the various photodiode geometries of the pixel set represented in *Figure 3-3*.

The curves share the same behavior, each displaying a pronounced maximum in the response. Pixel sets of different photodiode active-area geometrical shapes were tested, including circular, rectangular, and L-shaped. For each shape, a similar phenomenon was observed: each type of pixel gives a maximum response for particular illumination conditions. Note that for each wavelength (λ was changed from 450 nm to 650 nm) the measurements were performed under uniform light (an integrating sphere was used). The tradeoff between the two conflicting factors, integrated photocarriers and conversion gain, give rise to an optimum geometry and thus the maximum photoresponse.

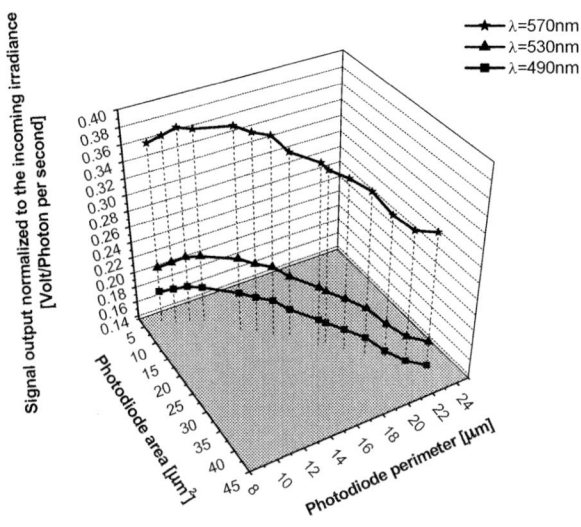

Figure 3-4. The signal output of the pixel set represented in *Figure 3-3* with illumination at three different wavelengths and with changes in the linear dimensions of the photodiodes.

A semi-analytical expression has been derived for a diffusion-limited pixel photosignal in a closely spaced photodiode array. In this expression each one of the parameters determining the signal output depends on the photodiode area and perimeter:

$$\frac{V_{out}(\lambda)}{N_{p\lambda}} = \left(\text{integration photocarriers} \times \text{conversion gain} \right)$$

$$= \frac{k_1 A + k_2 Pd\left(\dfrac{S-A}{S}\right)\left(1 - \dfrac{4Pi - P}{8L_{diff}}\right)}{k_3 A + k_4 P} \qquad (3.5)$$

The left part of this equation, $V_{out}(\lambda)/N_{p\lambda}$, corresponds to the pixel output voltage signal related to the number of incoming photons (in a time unit).

On the right side of equation (3.5), the denominator represents the conversion factor and the numerator represents the integrated photocarriers. The term in the denominator, $1/(k_3 A + k_4 P)$, represents the conversion gain, which depends on both the photodiode area and perimeter. The numerator

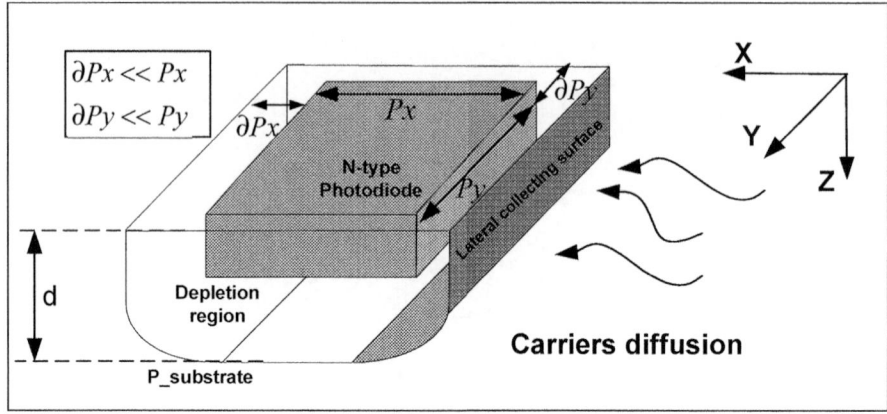

Figure 3-5. The sidewall or lateral collecting surface, built up from the junction bottom and sidewall depletion capacitances. P (in μm) is the photodiode perimeter; P_x and P_y are its dimensions in the x and y directions; d (in μm) is the junction depth; and ∂P_x and ∂P_y are the lateral depletion stretches.

consists of two terms that contribute to the total number of carriers collected by the imager. The first term, $k_1 A$, represents the contribution of the photocarriers created within the photodiode itself, i.e., the contribution of the active area. The second term

$$k_2 Pd\left(\frac{S-A}{S}\right)\left(1-\frac{4Pi-P}{8L_{diff}}\right)$$

represents the contribution of the periphery or the "lateral diffusion current", i.e., the carriers that had been created in the area surrounding the photodiode, had successfully diffused towards the photodiode, and had been collected.

Pd (in μm^2) represents the lateral collecting surface or the interaction cross-section for lateral diffusion, where P (in μm) is the photodiode perimeter, and d (in μm) is the junction depth. Since $\partial P_x << P_x$ and $\partial P_y << P_y$ (where ∂P_x and ∂P_y are the lateral depletion stretches of the photodiode), ∂P_x and ∂P_y can be neglected. Therefore, it can be assumed that the perimeter P (in μm) itself defines the boundary of lateral interaction area (see *Figure 3-5*).

The term $((S-A)/S)$ is dimensionless. Since the optical generation rate is relatively uniform throughout the substrate, this multiplier is proportional to the relative number of carriers created within the pixel substrate around the photodiode. Here, $(S-A)$ (in μm^2) represents the substrate area, i.e., the unoccupied area surrounding the photodiode within the pixel, whereas A

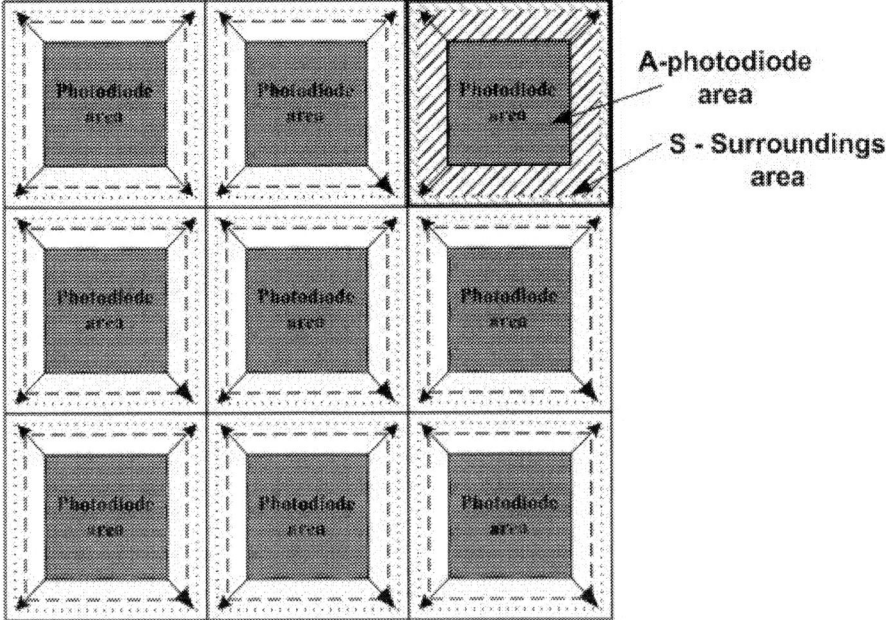

Figure 3-6. A schematic illustration of the boundary condition where the active area is increased to the maximum possible and reaches the pixel boundary (i.e., $A = S$ for all the pixels in the array). When this occurs, the diffusion contribution will equal zero.

(in μm^2) is the photodiode active area. The lower the fill factor, the higher the number of the carriers created within the photodiode surroundings that can diffuse and contribute to the total signal. It is clear that this multiplier represents the array influence on the pixel photodiode, and the boundary condition, i.e., as the active area increases to the pixel boundary and $A = S$ is approached for all the pixels in array, the diffusion contribution to the signal approaches zero. *Figure 3-6* illustrates the situation where the diffusion equals zero as a result of maximum expansion of the photodiode area for all pixels in the array. Therefore, the multiplier $((S - A)/S)$ represents the relative number of the potential diffusion contributors.

The term $(1 - (4Pi - P)/8L_{diff})$ is dimensionless and indicates the approximate relative distance that the prospective contributor has to pass before it is trapped in the depletion region. In this term, L_{diff} (in μm) is the characteristic diffusion length and Pi (in μm) is the pixel pitch. As the photodiode dimensions and the perimeter P (in μm) increase, the maximum distance that a carrier created within the substrate has to diffuse before it is collected by the peripheral sidewall collecting surface Pd decreases. Thus, the diffusion contribution increases. This multiplier is obtained from a series

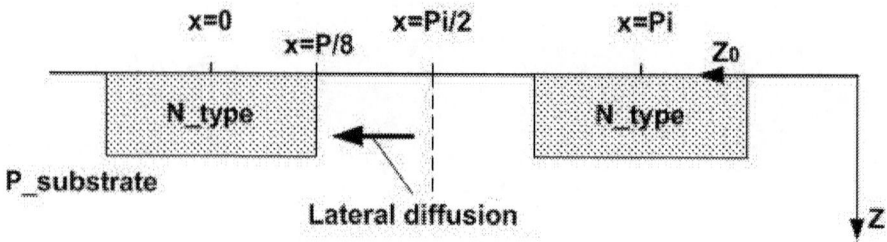

Figure 3-7. A schematic cross-section of two imager sites, illustrating the dependence of diffusion distance on the photodiode geometry. *Pi* is the pixel pitch.

expansion of the expression $exp[-(4Pi - P)/8L_{diff}]$, which represents the solution to the one-dimensional diffusion equation [19–21]. Since the distances between the photodiodes in the array—and therefore the maximum carrier path—are small compared to the minority carrier diffusion length, it is sufficient to consider only the first two terms of the expansion. *Figure 3-7* illustrates the dependence of diffusion distance on the photodiode geometry.

Also in equation (3.5), $V(\lambda)$ (in volts) is the pixel signal output for a particular wavelength.

$N_{p\lambda}$ (in photons/sec) is the photon irradiance. Since watts are joules per second, one watt of monochromatic radiation at wavelength λ corresponds to $N_{p\lambda}$ photons per second. The general expression is

$$\frac{dN_{p\lambda}}{dt} = 5.03 \times 10^{15} P_{\lambda} \lambda$$

where P_{λ} is in watts and λ is in nm.

The coefficient k_1 (in μm^{-2}) describes the unit active-area contribution to the total number of electrons collected by the imager, i.e., the number of electrons collected by the unit photodiode area in a time unit.

The coefficient k_2 (in μm^{-2}) describes the photodiode unit peripheral-area contribution to the total number of electrons collected by the imager, i.e., the number of electrons collected by the sidewall collecting surface within the substrate. *Figure 3-5* illustrates the sidewall or lateral collecting surface, which is built up from junction bottom and sidewall depletion capacitances.

The coefficients k_3 and k_4 (in $aF \cdot \mu m^{-2}$ and $aF \cdot \mu m^{-1}$, respectively) describe the bottom and sidewall capacitances in the regular junction capacitance sense (see equation (3.1)) and are defined by the particular process data.

In summary, the diffusion contribution to the overall signal is proportional to the lateral collecting area, the number of possible contributors, and the distance that the carrier has to pass before its collection by the photodiode. All parameters, with the exception of the coefficients k_1 and k_2, are defined by the process and design data. The expression agrees well with experimental results, as shown in the following section.

3.3 Comparison with experimental results

We have performed a functional analysis of experimental data obtained from a 14×14 μm pitch CMOS APS array by means of scientific analysis software. Pixel sets with square, rectangular, circular, and L-shaped active areas of different sizes have been tested. The response was analyzed for different wavelengths in the visible spectrum. Example results for a photodiode pixel with a square-shaped active area have been presented here.

The solution (in the minimum variance sense) of equation (3.5) for different pixel sets and different wavelengths enabled the extraction of the missing coefficients k_1 and k_2. The combination of these coefficients determined the contributions to the total pixel output signal and remained constant for all pixels at certain wavelength exposures.

It is evident from *Figure 3-4* that a longer wavelength enabled better response; for example, the signal obtained at 570 nm was almost two times the one obtained at 490 nm. This can be related to better absorption of red visible radiation within the semiconductor depth and thus better quantum efficiency. Moreover, the curves in *Figure 3-4* are shifted approximately in parallel. Since this output change was defined only by k_1 and k_2 (all other terms are wavelength independent), it was to be expected that the coefficients are wavelength dependent and that there was an increase in their values for longer wavelengths. As mentioned earlier, these coefficients represented the number of electrons collected by the photodiode via its upper and lateral faces respectively. Since the output signal was normalized to the number of photons impinging the pixel for uniform incident illumination, k_1 and k_2 was interpreted as the quantum efficiency per unit area for the upper and lateral faces respectively.

As the output signal for 570 nm was approximately double that for 490 nm, it was predicted that the QE for 570 nm would be double that the QE for 490 nm. The values obtained for the coefficients k_1 and k_2 confirmed this result: k_1/k_2 (570 nm) $\approx 0.468/0.229$, and k_1/k_2 (490 nm) $\approx 0.215/0.107$.

It is worth noting that QE, which is one of the main figures of merit for imagers, was considered only as a whole-pixel characteristic; i.e., no specific attention was paid to the photodiode shape and fill factor [1, 13, 22–23]. The work in this chapter demonstrates the merit of dividing QE into its main and diffusion portions according to the pixel geometrical shape and fill factor. This chapter also introduces a method (solving equation (3.5)) for the

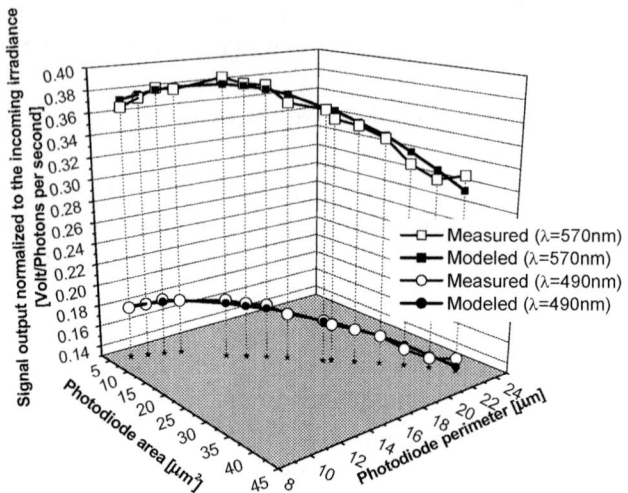

Figure 3-8. Comparisons at two different wavelengths of the derived expression, equation (3.5), with the actual pixel measurements presented in *Figure 3-3* and *Figure 3-4*.

determination and prediction of QE based on the process and design data. The result obtained for the presented pixel sets is $k_1/k_2 \approx 2.7$. This demonstrates that even though the active area is the primary source of the output, the substrate could account for up to 50% of the total output signal.

A 3-D graph presented in *Figure 3-8* shows example comparisons at two different wavelengths of the derived expression, equation (3.5), and the actual pixel measurements presented in *Figure 3-3* and in *Figure 3-4*. It illustrates the correspondence between the measured and the modeled output signal as the pixel area and perimeter are varied. Furthermore, the modeled function reaches its maximum exactly at the point marked by the measurements, confirming the proposed assumption that a compromise between integration photocarriers and conversion gain results in an optimum photodiode geometry. Thus, the model enables the successful prediction of a geometry that provides maximum output signal. Based on equation (3.5) and the specific process data, it is possible to select the photodiode shape and size that will provide the highest outcome.

The functional analysis of equation (3.5) enables the determination of the variables (area A and perimeter P) corresponding to the maximum value of the argument. It should be noted that area and perimeter are not independent; the function describing their dependence is implicit and must be taken into

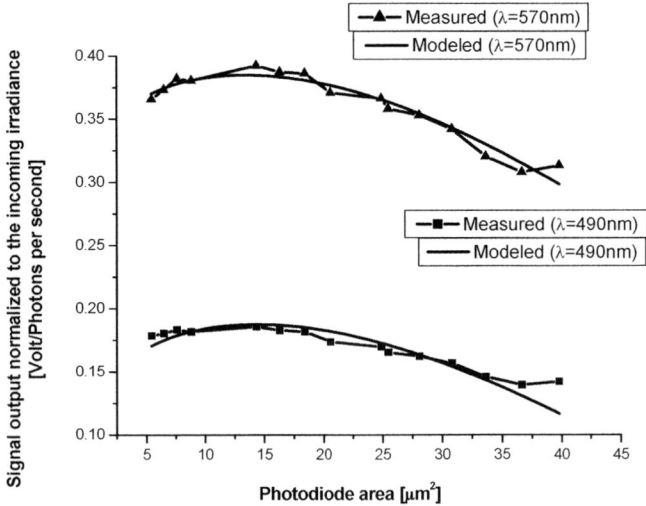

Figure 3-9. A comparison of the dependence of the modeled and experimental pixel output signal on the photodiode area for the square-shaped pixel data set of *Figure 3-8.*

consideration. In the case of a symmetrical photodiode shape, the function connecting area and perimeter can be easily obtained. equation (3.5) thus reduces to the case of only one independent variable; for example, the expression $P = 4A^{1/2}$ relates area and perimeter for the square-shaped photodiode. This result is confirmed in *Figure 3-9*, which presents a 2-D comparison of the dependence of the modeled and the experimental output signal on the photodiode area for the square-shaped pixel data set of *Figure 3-8*. A more complicated photodiode shape could always be represented as an aggregate of the elementary symmetrical parts [17, 24] and investigated in order to obtain a relation with only one independent variable as above.

Good correspondence of the model with the data is seen when the maximum divergence of the model result is constrained to 5%. It should be noted that surface leakage and the non-ideal transmission rate of the overlayers are not included in the present analysis and are considered to be second order effects [25].

3.4   CMOS APS pixel photoresponse prediction for scalable CMOS technologies

This section describes a semi-analytical diffusion-limited CMOS APS pixel photoresponse model use for maximum pixel photosignal prediction in scalable CMOS technologies.

3.4.1   Impact of technology scaling on sensitivity

Over the last twenty years, the evolution of CMOS technology has followed Moore's Law: a new generation of technology has been developed every three years, and between generations, memory capacity increased by a factor of four and logic circuit density increased by a factor of between two and three. Furthermore, every six years (two generations), the feature size decreased by a factor of two and transistor density, clock rate, chip area, chip power dissipation and the maximum number of pins has doubled. This continuous development has led to a reduction in the state-of-the-art commercially available minimum lithographic feature size from 3 μm in 1977 to 0.25 μm in 1998 and 0.1 μm in 2003. It is anticipated that it is technically possible to maintain the pace shown in *Table 3-1* [26].

CMOS process development is funded by high-volume sales of standard CMOS logic and memory chips. Hence, CMOS imaging technology does not have to bear the process development costs and consequently has cheaper process costs than CCD imagers.

Decreases in pixel size much beyond 5×5 μm have not been considered to be of much interest due to camera lens diffraction issues. However, in a common Bayer patterned RGB color sensor, 2×2 pixels define an effective color pixel; further downscaling of single pixels may prove useful for fitting an entire effective color pixel within the optical lens resolution limit. Therefore, pixel sizes below 5×5 μm^2 can be expected for all imagers, contrary to an assumption made by Wong [27].

The predicted downscaling of the supply voltage (see *Table 3-1*) will lead to a reduction of usable signal range and, together with increased noise components, will limit the overall dynamic range. Sources of detrimentally high leakage currents that generate shot noise and flood the pixel are discussed by Wong [27]. Examples include dark current, p-n junction tunneling current, transistor off current, gate oxide leakage, and hot carrier effects.

As Wong [27] noted, the diffusion collection behavior is modified by increasing doping levels through the reduction of mobility and lifetime. Under strong absorption conditions, the absorption coefficient $\alpha(\lambda)$ and the

Table 3-1. Technology parameters for different process generations (after [26]).

Process	0.8μm	0.5μm	0.35μm	0.25μm	0.18μm	0.1μm
Supply voltage (V)	5-3.3	5-3.3	5-3.3	2.5-1.8	1.8-1.5	1.2-0.9
Interconnect levels	2	2-3	4-5	5-6	6-7	8-9
Substrate doping (cm^{-3})	8×10^{16}	1.2×10^{17}	2.5×10^{17}	3.4×10^{17}	5×10^{17}	1×10^{18}
Junction depth d/S (nm)	350-450	300-400	200-300	50-100	36-72	20-40
Depletion region (μm)	0.71	0.57	0.39	0.24	0.19	0.1
Mobility (cm$^2\cdot$V$^{-1}\cdot$s^{-1})	825	715	550	485	425	345
Lifetime (μs)	3.6	2.3	1.1	0.8	0.6	0.3
Diffusion length (μm)	88	68	41	33	25	15

diffusion length L_n are both much smaller than the width of the substrate. Further effects on spectral sensitivity result from the shallower drain/source diffusion and the shrinkage of the depletion widths. Figure 3-10 demonstrates that the scaling effect on diffusion collection efficiency works in favor of CMOS imagers. The influence on mobility and lifetime are not very strong, so the loss of sensitivity in the infrared range is not very pronounced. On the other hand, the shallower diffusions promote the collection of blue light in the silicon material. This improves performance for visible light applications, especially for color sensors, as was shown above. These effects of technology scaling apply directly to the photodiode. However, the losses caused by the overlying layers must also be included and reduce the total QE considerably.

The lens diffraction limit of 5 μm has been reached with the 0.35 μm process. The increased number of routing levels leads to an increasing

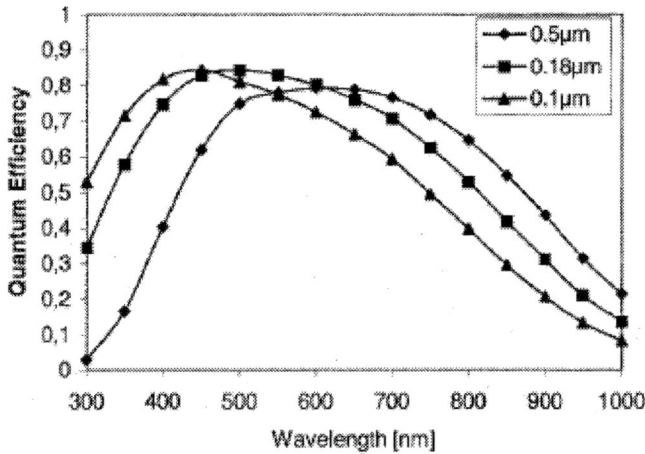

Figure 3-10. Internal carrier collection efficiency for different process generations (after [26]).

Table 3-2. Scaling trends of CMOS sensor parameters for a standard photodiode (after [26]).

Technology	0.8 µm	0.35 µm	0.25 µm	0.18 µm
Sensor type	standard	standard	standard	standard
A_{pix} (µm²)	14×14	7×7	5×5	5×5
Fill factor (%)	60	60	60	80
A_{diode} (µm²)	117.6	29.4	15	20
C_{diff} (fF)	5.75	2.7	1.87	0.65
C_{gate} (fF)	2.06	1.1	0.67	0.48
C_{diode} (fF)	94.4	37.8	38.3	24.2
C_{tot} (fF)	102.2	41.6	40.8	25.3
q/C (µV/e⁻)	1.5	3.8	3.9	6.4
S (V·µJ⁻¹·cm⁻²)	1.8	1.1	0.57	1.2

thickness of the layer system on top of the diode, which aggravates the problems of absorption, reflection, diffraction, and even light guidance into distant pixels. In addition, the accuracy requirements for microlenses become very strict as the sensitive spot in the pixels becomes smaller and moves further away from the lens. Below 0.35 µm, only those processes that allow blocking of silicide layers on diffusions can be employed. The shrinkage of capacitances does not generally compensate sufficiently for the reduction of sensitive area. On the contrary, the area-specific capacitance increases due to increased doping concentrations, which automatically reduces the sensitivity. For a given fill factor, the overall sensitivity remains low, as shown by the results presented in *Table 3-2* [26].

3.4.2   CMOS APS pixel photoresponse predictions for scalable CMOS technologies

It has recently been shown ([28] and section 3.2) that for any pixel active-area shape, a reliable estimate of the degradation of image performance is possible. The tradeoff between conflicting factors (such as integrated photocarriers and conversion gain) can be compared for each pixel design, allowing the optimum overall sensor performance in a particular technology process to be determined. The present work is based on a thorough study of the experimental data acquired from several pixel chips fabricated in two different technology processes, 0.5 µm and 0.35 µm CMOS processes. Using this data, the analysis is extended to show the efficacy and the suitability of this photoresponse model in scalable CMOS processes. Calculations based on this model only require readily available process and design data to make possible the selection of designs with maximum output signal.

Figure 3-3 and *Figure 3-11* show two subsets of pixels (square and rectangular active-area shape, respectively) with decreasing photodiode dimensions and fabricated in a standard CMOS 0.5 µm process.

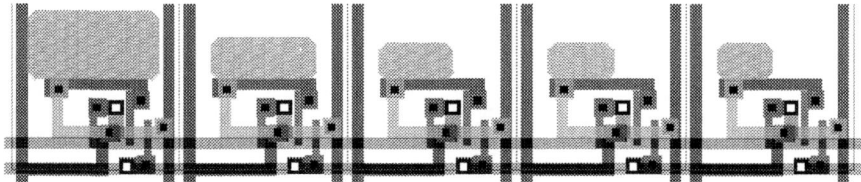

Figure 3-11. A sample from an otherwise identical set of rectangular active-area pixels (CMOS 0.5 µm technology) with decreasing photodiode dimensions. The photodiode areas in the full pixel set vary from 63 µm² to 13 µm² and their perimeters from 34 µm to 15.5 µm.

Figure 3-4 and *Figure 3-12* show the corresponding output curves for several wavelengths of illumination. These curves share the same behavior; i.e., *Figure 3-4* curves display a pronounced maximum response location, while in *Figure 3-12* the curves tend to an extremum.

The photoresponse model enables the extraction of the unit "main area" and unit peripheral contributions to the output signal, and in turn the identification and modeling of pixel behavior (see *Figure 3-8*).

The fact that the combination of these two contributions remains invariable for all pixels at a specific wavelength of illumination (for a

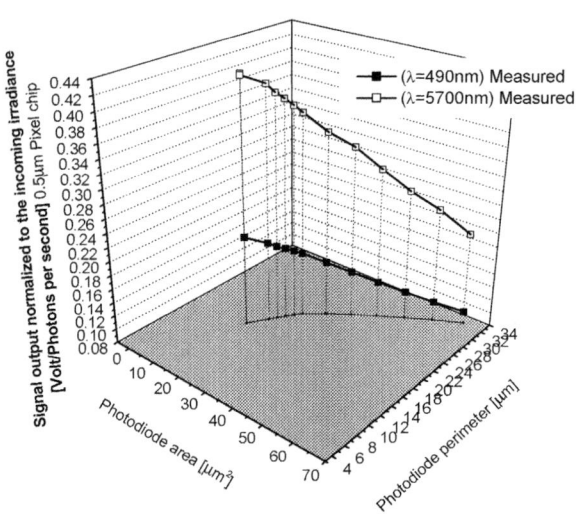

Figure 3-12. Measured signal output obtained for the pixel set presented in *Figure 3-11* as functions of the photodiode linear dimensions for two different wavelength illuminations.

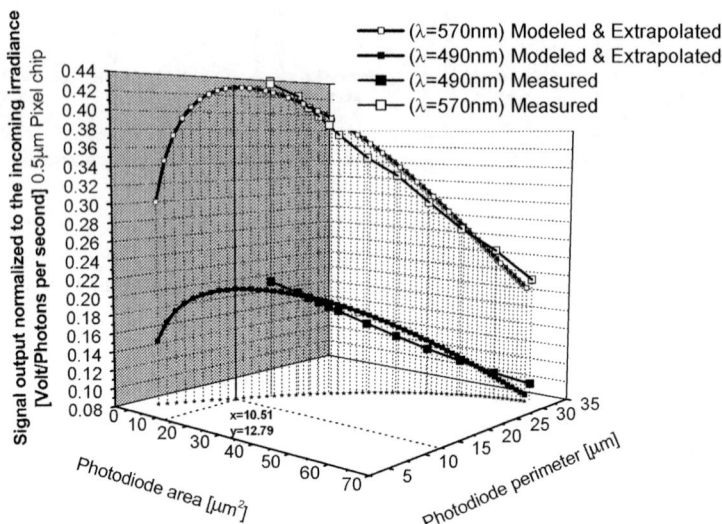

Figure 3-13. A comparison of the modeled results and the measured data at two different wavelengths for the pixel set represented in *Figure 3-11*. Extrapolation of the model predicted a photodiode area and perimeter corresponding to a pixel design that would enable maximum photoresponse. Note that the optimal pixel was not actually designed and measured, but rather its existence and geometric dimensions were revealed by extrapolation of the model.

specific process) enables the extrapolation of the modeled function and in turn the identification of the optimal photodiode geometry (for the rectangular pixel case, see *Figure 3-13*). Thus, the model theoretically predicts both the existence and the properties of the optimal geometry; (the optimal dimensions) based on the investigated process and the specific design data.

The total "main area" and total periphery contributions to the output signal have been examined separately as a function of the change in the photodiode dimensions. With a increase in the dimensions, the "main area" contribution drops and the periphery contribution rises such that they intercept. The interception point is at the exact location where the maximum output signal was predicted by extrapolation in *Figure 3-13* for the particular scalable CMOS process (see *Figure 3-14*).

The overall influence of scaling on the device sensitivity is very complicated and depends on a large variety of parameters. An analytical expression that uniquely determines the general scaling trends has not yet been developed [26, 27]. An approximation is proposed to describe the scaling influence: it is assumed that the ratio between the unit "main area" and the unit "periphery" contributions has a slight upward trend, due

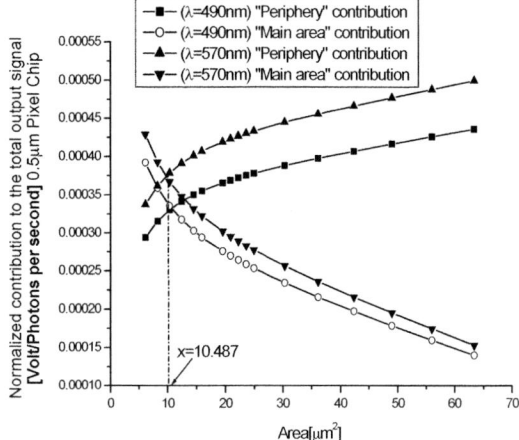

Figure 3-14. Total "main area" and periphery contributions to the output signal as a function of the change in the (rectangular) photodiode dimensions for the 0.5 μm CMOS pixel set presented in *Figure 3-11*. The interception point is at the exact location where the maximum output signal was predicted by extrapolation in *Figure 3-13*.

mostly to the reduction of mobility and lifetime with increasing doping levels and to shrinkage of the depletion widths. Under strong absorption conditions where the diffusion coefficient $\alpha(\lambda)$ and the diffusion length L_n are both much smaller than the substrate width, the photocurrent density due to diffusion in the substrate can be derived [26] as $J_{ph} = q\,\Phi/(1 + 1/\alpha(\lambda)\,L_n)$, where Φ is the photon flux entering the quasi-neutral p-substrate region and q is the electron charge. With technology downscaling, the unit "periphery" contribution to the output signal decreases. The depletion width shrinkage means that carriers are collected more through the bottom facet of the depletion region rather than through its lateral facets, intensifying therefore the relative "main area" contribution. In addition, the junction depth for advanced processes is small in comparison to the absorption depth, such that most photocarriers are collected through the bottom depletion facet. Based on the assumption above and using the process data and the extracted results (the coefficients k_1 and k_2, i.e., determining the unit "main area" and the unit "periphery" contributions for each particular wavelength), it is possible to determine the coefficients k_1 and k_2 for the more advanced scalable CMOS technology. It is predicted that

$$
k_1 / k_2 \Big|_{CMOS\ 0.35\mu m}^{570nm} \approx k_1 / k_2 \Big|_{CMOS\ 0.5\mu m}^{570nm}
$$
$$
\times \left(d_{CMOS\ 0.5\mu m} / d_{CMOS\ 0.35\mu m} \right) \cdot \left(Ln_{CMOS\ 0.35\mu m} / Ln_{CMOS\ 0.5\mu m} \right) \approx 1.12 \tag{3.6}
$$

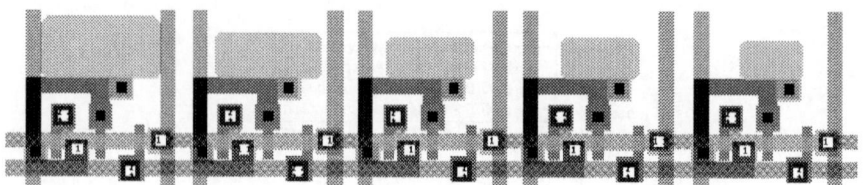

Figure 3-15. A subset of a rectangular-shaped active area pixel set with decreasing photodiode dimensions (CMOS 0.35 μm technology). The photodiode areas vary from 13.4 μm² down to 4.3 μm² and their perimeters vary from 15 μm down to 8.1 μm.

and

$$k_1 / k_2\Big|_{CMOS\ 0.35\mu m}^{490nm} \approx k_1 / k_2\Big|_{CMOS\ 0.5\mu m}^{490nm}$$
$$\times \left(d_{CMOS\ 0.5\mu m} / d_{CMOS\ 0.35\mu m}\right) \cdot \left(Ln_{CMOS\ 0.35\mu m} / Ln_{CMOS\ 0.5\mu m}\right) \approx 1.07 \tag{3.7}$$

where d is the depletion depth.

An example calculation was performed for set of pixels of a rectangular photodiode shape designed according to the CMOS 0.35 μm rules and obeying the same mathematical guidelines as the older technology of pixels presented in *Figure 3-12*. An example subset of these newer pixels is shown in *Figure 3-15*. All the pixels share a common traditional three-transistor-type readout circuitry, enabling identification of the behavior of different pixel types.

The predictions of equations (3.6) and (3.7) were used to find a pixel that enabled maximum response. In *Figure 3-16*, the interception point of the calculated total "main area" and the total "periphery" contributions envisaged the maximum photoresponse pixel geometry for the pixel set designed and fabricated in the more advanced CMOS 0.35 μm technology design (shown in *Figure 3-15*).

Figure 3-17 shows the comparison between the measured and theoretically modeled output curves for several wavelengths of illumination where obvious maximum response geometry was indicated. Note that the modeled function (based on available process and design data) reaches its maximum exactly at the point marked by the measurements. Moreover, the values obtained from the measurements of the contributions ratio are

$$k_1 / k_2\Big|_{CMOS\ 0.35\mu m}^{570nm} \approx 1.13$$

$$\tag{3.8}$$

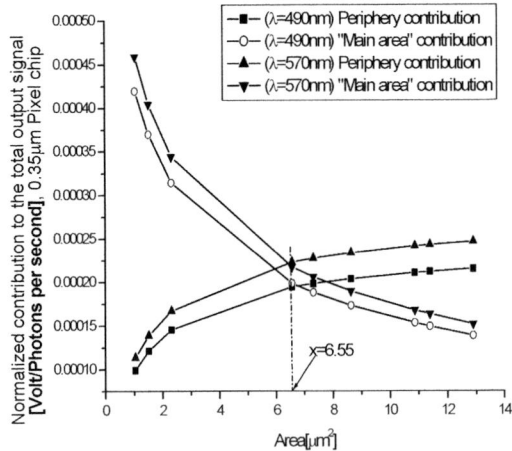

Figure 3-16. Total "main area" and periphery contributions to the output signal as a function of the changes in the photodiode dimensions for the 0.35 μm CMOS pixel set presented in *Figure 3-15*. Note that this result is obtained theoretically, based only on the experimental results obtained from an older CMOS process and scaling considerations.

and

$$k_1 / k_2 \Big|_{CMOS\ 0.35\mu m}^{490nm} \approx 1.068 \tag{3.9}$$

These are similar to our theoretical results in equations (3.6) and (3.7). The maximum occurs exactly at the previously predicted interception point (*Figure 3-16*).

The theoretical result described here was obtained for the 0.35 μm CMOS design based only on the parameters extracted from the available design and process data; i.e., there was no need for an actual study with a 0.35 μm test chip. The optimum geometry, or the pixel that enabled the maximum photoresponse, was predicted theoretically based only on the experimental results obtained from older process data and scaling considerations. The measurements confirmed the theoretical prediction.

This model for photoresponse estimation of a photodiode based CMOS APS field of applicability does not appear to be constrained by a specific technology process. It can therefore be used as a predictive tool for design optimization of each potential application in scalable CMOS technologies.

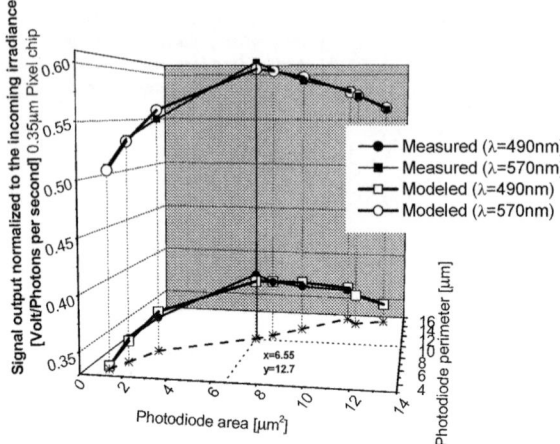

Figure 3-17. A comparison of the modeled and the measured results obtained for the pixels presented in *Figure 3-15* at two different wavelengths of illumination. The geometry of the pixel that enabled maximum photoresponse is indicated. The maximum occurred exactly at the interception point indicated in *Figure 3-16*; i.e., the theoretically predicted result was confirmed by the measurements.

3.5   Summary

A closed-form semi-analytical expression has been presented for diffusion-limited CMOS APS pixel photosignal in a closely spaced photodiode array. This expression represents the pixel photosignal dependence on the pixel geometrical shape and fill factor, i.e., the photodiode active area and perimeter. It enables identification of the behavior of different pixel types and shows how changes in the device geometry affect its overall performance. In addition, the expression introduces a method for the determination and prediction of CMOS photodiode quantum efficiency based on the process and design data.

The results indicate that the tradeoffs between conflicting factors (such as integration photocarriers and conversion gain) can be compared for each potential pixel design, and a reliable estimate for optimum overall sensor performance is possible.

The model clearly makes possible the theoretical prediction of pixel designs that enable the extraction of maximum output signal for any selected photodiode shape. This prediction is based only on the usual process and design data available for different scalable CMOS technologies, and thus can be a practical tool for design optimization.

BIBLIOGRAPHY

[1] O. Yadid-Pecht, B. Mansoorian, E. Fossum, and B. Pain, "Optimization of active pixel sensor noise and responsivity for scientific applications," in *Proc. IS&T/SPIE Symp. Electronic Imaging: Science and Technology*, San Jose, California, Feb. 1997.

[2] O. Yadid-Pecht, R. Ginosar, and Y. Shacham-Diamand, "A random access photodiode array for intelligent image capture," *IEEE Trans. Electron Devices*, vol. 38, no. 8, pp. 1772–1781, Aug. 1991.

[3] E. Fossum, "Digital camera system on a chip," *IEEE Micro*, vol. 18, no. 3, pp. 8–15, May–June 1998.

[4] H. Wong, "Technology and device scaling considerations for CMOS imagers," *IEEE Trans. Electron Devices*, vol. 43, no. 12, pp. 2131–2142, Dec. 1996.

[5] O. Yadid-Pecht, "CMOS imagers," course notes, Ben-Gurion University, Beer-Sheva, Israel, 2000.

[6] A. Moini, "Vision chips or seeing silicon: technical report," Centre for High Performance Integrated Technologies and Systems, University of Adelaide, Mar. 1997. Available: www.eleceng.adelaide.edu.au/Groups/GAAS/Bugeye/visionchips/index.html

[7] J. Hynecek, "BCMD — An improved photosite structure for high-density image sensors," *IEEE Trans. Electron Devices*, vol. 38, ED-5, pp. 1011–1020, May 1991.

[8] K. Matsumoto, I. Takayanagi, T. Nakamura, and R. Ohta, "The operation mechanism of a charge modulation device (CMD) image sensor," *IEEE Trans. Electron Devices*, vol. 38, ED-5, pp. 989–998, May 1991.

[9] S. Mendis, S. Kemeny, R. Gee, B. Pain, C. Staller, Q. Kim, and E. Fossum, "CMOS active pixel image sensors for highly integrated imaging systems," *IEEE J. Solid State Circuits*, vol. 32, pp. 187–197, Feb. 1997.

[10] P. Magnan, A. Gautrand, Y. Degerli, C. Marques, F. Lavernhe, C. Cavadore, F. Corbiere, J. Farre, O. Saint-Pe, M. Tulet, and R. Davancens, "Influence of pixel topology on performances of CMOS APS imagers," *Proc. SPIE*, vol. 3965, 2000.

[11] J. Bogaerts and B. Dierickx, "Total dose effects on CMOS active pixel sensors," *Proc. SPIE*, vol. 3965, pp. 157–167, 2000.

[12] H. Tian, B. Fowler, and A. El-Gamal, "Analysis of temporal noise in CMOS photodiode active pixel sensor," *IEEE J. Solid State Circuits*, vol. 36, pp. 92–100, Jan. 2001.

[13] H. Tian, X. Liu, S.H. Lim, S. Kleinfelder, and A. El-Gamal, "Active pixel sensors fabricated in a standard 0.18 μm CMOS technology," *Proc. SPIE*, vol. 4306, pp. 441–449, 2001.

[14] J. P. Lavine, E. A. Trabka, B. C. Burkey, T. J. Tredwell, E. T. Nelson, and C. N. Anagnosyopoulos, "Steady-state photocarrier collection in silicon imaging devices," *IEEE Trans. Electron Devices*, vol. 30, ED-9, pp. 1123–1134, Sept. 1983.

[15] D. Kavaldjiev and Z. Ninkov, "Subpixel sensitivity map for a charge coupled device sensor," *Opt. Eng.*, vol. 37, no. 3, pp. 948–954, Mar. 1998.

[16] J. S. Lee and R. I. Hornsey, "Photoresponse of photodiode arrays for solid-state image sensors," *J. Vacuum Sci. Technol.*, vol. 18, no. 2, pp. 621–625, Mar. 2000.

[17] I. Shcherback and O. Yadid-Pecht, "CMOS APS MTF modeling," *IEEE Trans. Electron Devices*, vol. 48, ED-12, pp. 2710–2715, Dec. 2001.

[18] J. Tandon, D. Roulston, and S. Chamberlain, "Reverse-bias characteristics of a P^+-N-N^+ photodiode," *Solid-State Electron.*, vol. 15, pp. 669–685, Jun. 1972.

[19] S. G. Chamberlain and D. H. Harper, "MTF simulation including transmittance effects of CCD," *IEEE Trans. Electron Devices*, vol. 25, ED-2, pp. 145–154, 1978.

[20] R. Stern, L. Shing, and M. Blouke, "Quantum efficiency measurements and modeling of ion-implanted, laser-annealed charge-coupled devices: X-ray, extreme ultraviolet, ultraviolet, and optical data," *Appl. Opt.*, vol. 33, no. 13, pp. 2521–2533, May 1994.

[21] I. Shcherback and O. Yadid-Pecht, "CMOS APS MTF Modeling," *IEEE Trans. Electron Devices*, vol. 48, ED-12, pp. 2710–2715, Dec. 2001.

[22] B. Fowler, A. El-Gamal, D. Yang, and H. Tian, "A method for estimating quantum efficiency for CMOS image sensors," *Proc. SPIE*, vol. 3301, pp. 178–185, 1998.

[23] P. B. Catrysse, X. Liu, and A. El-Gamal, "QE reduction due to pixel vignetting in CMOS image sensors," *Proc. SPIE*, vol. 3965, pp. 420–430, 2000.

[24] O. Yadid-Pecht, "The geometrical modulation transfer function (MTF) for different pixel active area shapes," *Opt. Eng.*, vol. 39, no. 4, pp. 859–865, 2000.

[25] D. Ramey and J. T. Boyd, "Computer simulation of optical crosstalk in linear imaging arrays," *IEEE J. Quantum Electron.*, vol. 17, pp. 553–556, Apr. 1981.

[26] T. Lule, S. Benthien, H. Keller, F. Mutze, P. Rieve, K. Seibel, M. Sommer, and M. Bohm, "Sensitivity of CMOS-based imagers and scaling perspectives," *IEEE Trans. Electron Devices*, vol. 47, ED-11, pp. 2710–2722, Nov. 2000.

[27] H. Wong, "Technology and device scaling considerations for CMOS imagers," *IEEE Trans. Electron Devices*, vol. 43, no. 12, pp. 2131–2142, Dec. 1996.

[28] I. Shcherback and O. Yadid-Pecht, "Photoresponse analysis and pixel shape optimization for CMOS active pixel sensors," *IEEE Trans. Electron Devices* (Special Issue on Image Sensors), vol. 50, pp. 12-19, Jan. 2003.

Chapter 4

ACTIVE PIXEL SENSOR DESIGN: FROM PIXELS TO SYSTEMS

Alexander Fish and Orly Yadid-Pecht
The VLSI Systems Center
Ben-Gurion University
P.O.B. 653 Beer-Sheva 84105, ISRAEL

Abstract: Since active pixel sensors (APS) are fabricated in a commonly used CMOS process, image sensors with integrated "intelligence" can be designed. These sensors are very useful in many scientific, commercial and consumer applications. Current state-of-the-art CMOS imagers allow integration of all functions required for timing, exposure control, color processing, image enhancement, image compression, and ADC on the same die. In addition, CMOS imagers offer significant advantages and rival traditional charge coupled devices (CCDs) in terms of low power, low voltage and monolithic integration. This chapter presents different types of CMOS pixels and introduces the system-on-a-chip approach, showing examples of two "smart" APS imagers. The camera-on-a-chip approach is introduced, focusing on the advantages of CMOS sensors on CCDs. Different types of image sensors are described and their modes of operation briefly explained. Two examples of CMOS imagers are presented, a smart vision system-on-a-chip and a smart tracking sensor. The former is based on a photodiode APS with linear output over a wide dynamic range, made possible by random access to each pixel and by the insertion of additional circuitry into the pixels. The latter is a smart tracking sensor employing analog non-linear winner-take-all (WTA) selection.

Key words: CMOS image sensor, active pixel sensor (APS), charge-coupled devices (CCD), passive pixel (PS), system-on-a-chip, smart sensor, dynamic range (DR), winner-take-all (WTA) circuit.

4.1   Introduction

Driven by the demands of multimedia applications, image sensors have become a major category of high-volume semiconductor production. The introduction of imaging devices is imminent in consumer applications such as cell phones, automobiles, computer-based video, smart toys and both still and video digital cameras.

In addition to image capture, the electronics in a digital camera must handle analog-to-digital (ADC) conversion as well as a significant amount of digital processing for color imaging, image enhancement, compression control and interfacing. These functions are usually implemented with many chips fabricated in different process technologies.

The continuous advances in CMOS technology for processors and DRAMs have made CMOS sensor arrays a viable alternative to the popular charge-coupled devices (CCD) sensor technology. New technologies provide the potential for integrating all imaging and processing functions onto a single chip, greatly reducing the cost, power consumption and size of the camera [1–3]. Standard CMOS mixed-signal technology allows the manufacture of monolithically integrated imaging devices: all the functions for timing, exposure control and ADC can be implemented on one piece of silicon, enabling the production of the so-called "camera-on-a-chip" [4]. *Figure 4-1* is a diagram of a typical digital camera system, showing the difference between the building blocks of commonly used CCD cameras and the CMOS camera-on-a-chip. The traditional imaging pipeline functions— such as color processing, image enhancement and image compression—can also be integrated into the camera. This enables quick processing and exchanging of images. The unique features of CMOS digital cameras allow many new applications, including network teleconferencing, videophones, guidance and navigation, automotive imaging systems, and robotic and machine vision.

Most digital cameras currently use CCDs to implement the image sensor. State-of-the-art CCD imagers are based on a mature technology and present excellent performance and image quality. They are still unsurpassed for high sensitivity and long exposure time, thanks to extremely low noise, high quantum efficiency and very high fill factors. Unfortunately, CCDs need specialized clock drivers that must provide clocking signals with relatively large amplitudes (up to 10 V) and well-defined shapes. Multiple supply and bias voltages at non-standard values (up to 15 V) are often necessary, resulting in very complex systems.

Figure 4-2 is a block diagram of a widely used interline transfer CCD image sensor. In such sensors, incident photons are converted to charge, which is accumulated by the photodetectors during exposure time. In the subsequent readout time, the accumulated charge is sequentially transferred into the vertical and horizontal CCDs and then shifted to the chip-level output amplifier. However, the sequential readout of pixel charge limits the readout speed. Furthermore, CCDs are high-capacitance devices and during

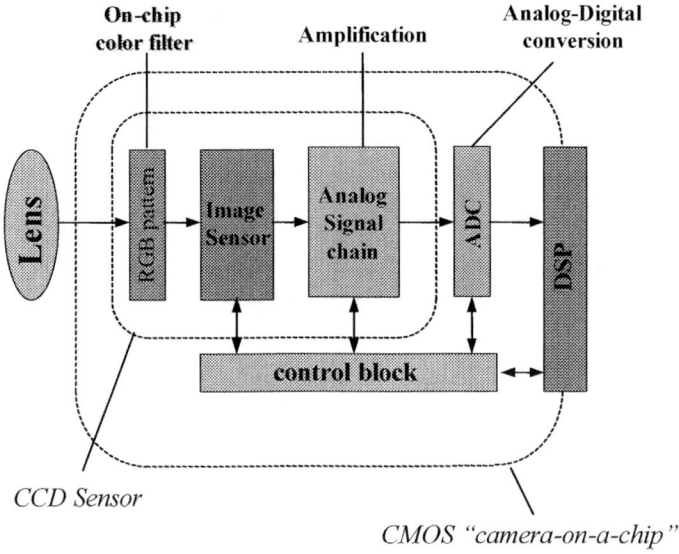

Figure 4-1. Block diagram of a typical digital camera system.

readout, all the capacitors are switched at the same time with high voltages; as a result, CCD image sensors usually consume a great deal of power. CCDs also cannot easily be integrated with CMOS circuits due to additional fabrication complexity and increased cost. Because it is very difficult to integrate all camera functions onto a single CCD chip, multiple chips must be used. A regular digital camera based on CCD image sensors is therefore burdened with high power consumption, large size and a relatively complex design; consequently, it is not well suited for portable imaging applications.

Unlike CCD image sensors, CMOS imagers use digital memory style readout, using row decoders and column amplifiers. This readout overcomes many of the problems found with CCD image sensors: readout can be very fast, it can consume very little power, and random access of pixel values is possible so that selective readout of windows of interest is allowed. *Figure 4-3* shows the block diagram of a typical CMOS image sensor.

The power consumption of the overall system can be reduced because many of the supporting external electronic components required by a CCD sensor can be fabricated directly inside a CMOS sensor. Low power consumption helps to reduce the temperature (or the temperature gradient) of both the sensor and the camera head, leading to improved performance.

An additional advantage of CMOS imagers is that analog signal processing can be integrated onto the same substrate; this has already been demonstrated by some video camera-on-a-chip systems. Analog signal

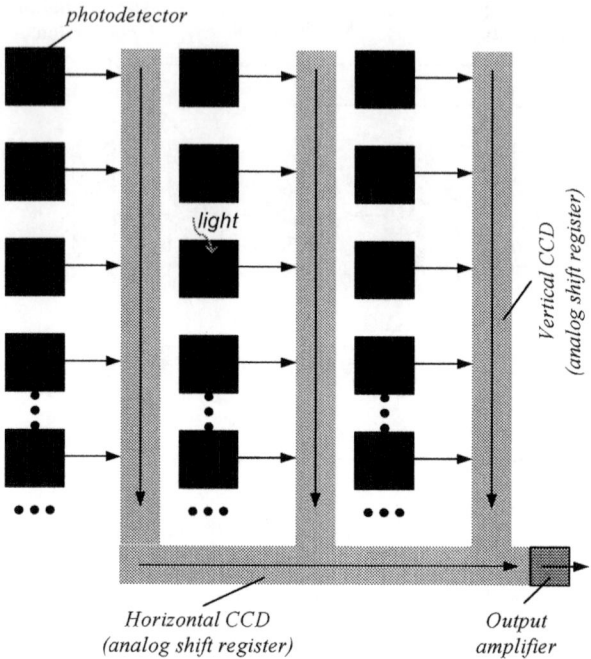

Figure 4-2. Block diagram of a typical interline transfer CCD image sensor.

processing can include widening the dynamic range of the sensor, real-time object tracking, edge detection, motion detection and image compression. These functions are usually performed by nonlinear analog circuits and can be implemented inside the pixels and in the periphery of the array. Offloading signal processing functions makes more memory and DSP processing time available for higher-level tasks, such as image segmentation or tasks unrelated to imaging.

This chapter presents a variety of implementations of CMOS image sensors, focusing on two examples of system-on-a-chip design: an image sensor with wide dynamic range (DR) [5] and a tracking CMOS imager employing analog winner-take-all (WTA) selection [6].

4.2   CMOS image sensors

CMOS pixels can be divided into two main groups, passive pixel sensors (PPS) and active pixel sensors (APS).

Each individual pixel of a PPS array has only a photosensing element (usually a photodiode) and a switching MOSFET. The signal is detected

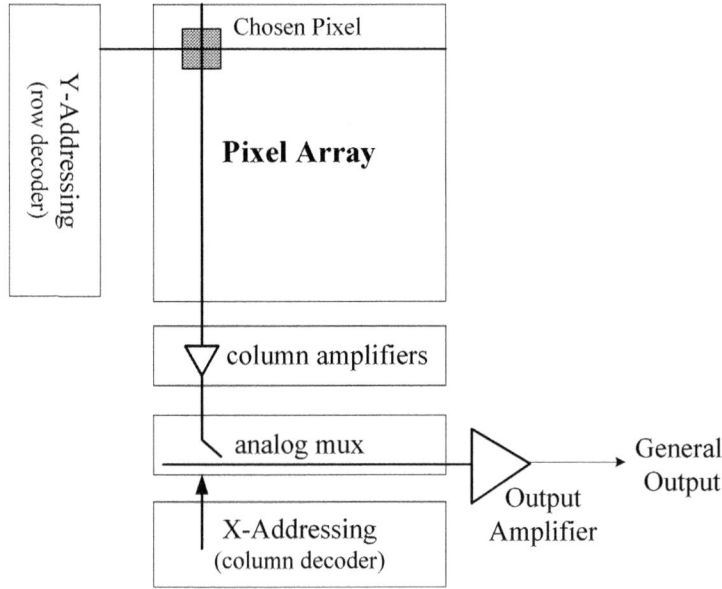

Figure 4-3. Block diagram of a typical CMOS image sensor.

either by an output amplifier implemented in each column or by a single output for the entire imaging device. These conventional MOS-array sensors operate like an analog DRAM, offering the advantage of random access to the individual pixels. They suffer from relatively poor noise performance and reduced sensitivity compared to state-of-the-art CCD sensors.

APS arrays are novel image sensors that have amplifiers implemented in every pixel; this significantly improves the noise parameter.

4.2.1 Passive Pixel Sensors

The PPS consists of a photodiode and just one transistor (labeled *TX* in *Figure 4-4*). *TX* is used as a charge gate, switching the contents of the pixel to the charge integration amplifier (CIA). These passive pixel CMOS sensors operate like analog DRAMs, as shown in *Figure 4-5*.

More modern PPS implementations use a CIA for each column in the array, as shown in *Figure 4-6*. The CIA readout circuit is located at the bottom of each column bus (to keep the voltage on that bus constant) and uses just one addressing transistor. The voltage V_{ref} is used to reset the photo-diode to reverse bias. Following the reset, this switch is opened, for a period of integration time (T_{int}). During this period, the photodiode discharges at a rate approximately proportional to the amount of incident

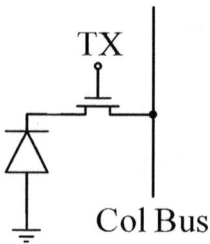

Figure 4-4. Passive pixel sensor structure.

illumination. When the MOS switch is closed again to reset the photodiode once more, a current flows via the resistance and capacitance of the column bus due to the difference between V_{ref} and the voltage on the diode (V_{diode}). The total charge that flows to reset the pixel is equal to that discharged during the integration period. This charge is integrated on the capacitor C_{int} and output as a voltage. When the final bus and diode voltages return to V_{ref} via the charge amplifier, the address MOS switch is turned off, the voltage across C_{int} is removed by the *Reset* transistor, and the integration process starts again.

The passive pixel structure has major problems due to its large capacitive loads. Since the large bus is directly connected to each pixel during readout,

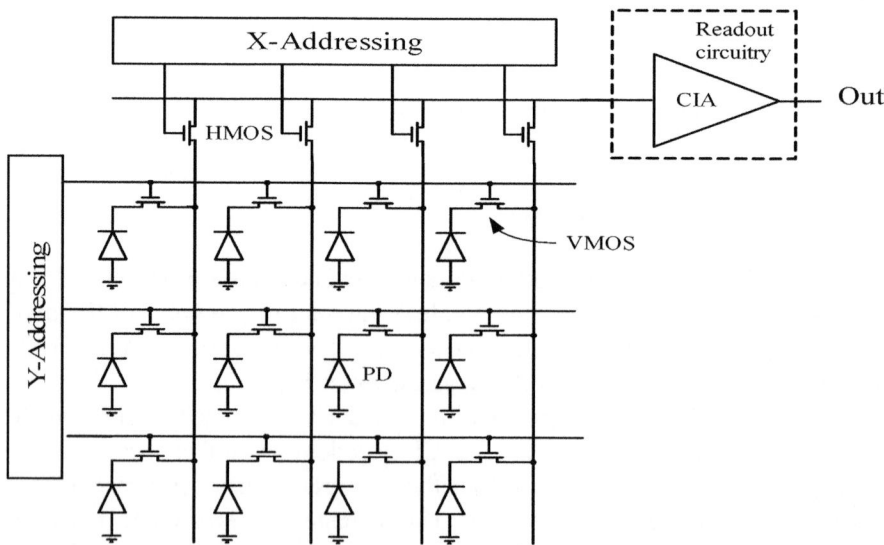

Figure 4-5. Basic PPS architecture.

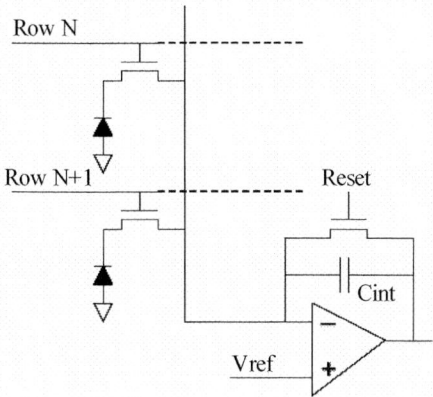

Figure 4-6. PPS implementation with a separate CIA for each column in the array (after [13]).

the *RC* time constant is very high and the readout is slow. In addition, passive pixel readout noise is typically high—on the order of 250 electrons rms compared to less than 10 electrons rms for commercial CCDs. Because of these factors, PPS does not scale well to larger array sizes or faster pixel readout rates. Furthermore, differences between the individual amplifiers at the bottoms of the different columns will cause fixed pattern noise (FPN). FPN is time-independent and arises from component mismatch due to variations in lithography, doping and other manufacturing processes.

PPS also offers advantages. For a given pixel size, it has the highest design fill factor (the ratio of the light sensitive area of a pixel to its total area), since each pixel has only one transistor. In addition, its quantum efficiency (QE)—the ratio between the number of generated electrons and the number of impinging photons—can be quite high due to this large fill factor.

4.2.2 Active Pixel Sensors

The passive pixel sensor was introduced by Weckler in 1967 [7]. The problems of PPS were recognized, and consequently a sensor with an active amplifier (a source follower transistor) within each pixel was proposed [8]. The current term for this technology, active pixel sensor, was first introduced by Fossum in 1992 [1]. *Figure 4-7* shows the general architecture of an APS array and the principal pixel structure. A detailed description of the readout procedure will be presented later in this chapter.

Active pixels typically have a fill factor of only 50–70%, which reduces the photon-generated signal. However, the reduced capacitance in each pixel

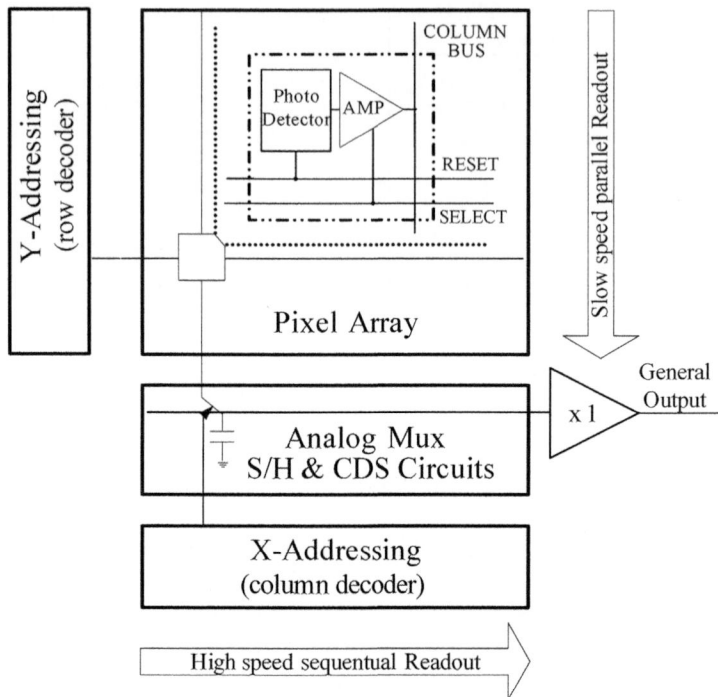

Figure 4-7. General architecture of an APS array.

leads to lower read noise for the array, which increases both the dynamic range and the signal-to-noise ratio (SNR).

The pixels used in these sensors can be divided into three types: photodiodes, photogates, and pinned photodiodes. The most popular is currently the photodiode.

4.2.3   Photodiode APS

The photodiode APS was described by Noble in 1968 [8] and has been under investigation by Andoh since the late 1980s [9]. A novel technique for random access and electronic shuttering with this type of pixel was proposed by Yadid-Pecht in the early 1990s [12].

The basic photodiode APS employs a photodiode and a readout circuit of three transistors: a photodiode reset transistor (*Reset*), a row select transistor (*RS*) and a source-follower transistor (*SF*). The scheme of this pixel is shown in *Figure 4-8*.

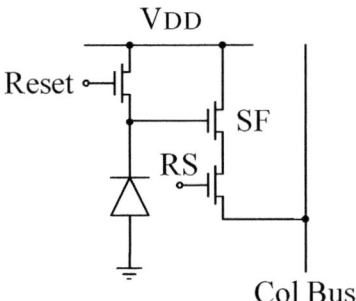

Figure 4-8. Basic photodiode APS pixel.

The charge-to-voltage conversion occurs at the sense node capacitance, which comprises the photodiode capacitance and all other parasitic capacitances connected to that node. In this case, these are the source capacitance of the *Reset* transistor and the gate capacitance of the *SF* transistor. The *SF* transistor acts as a buffer amplifier to isolate the sensing node; the load of this buffer (the active-current-source load) is located on each column rather than on each pixel to keep the fill factor high and to reduce pixel-to-pixel variations. The *Reset* transistor controls an integration time and is usually implemented with an NMOS transistor. Since no additional well is required for NMOS implementation, this allows a higher fill factor. However, an NMOS transistor with V_{DD} on both gate and drain can only reach a source voltage (at the photodiode node) of $V_{DD} - V_T$, thereby decreasing the dynamic range of the pixel. An example of a mask layout for this pixel architecture is shown in *Figure 4-9*.

Photodiode APS operation and readout are described here with reference to both *Figure 4-7* and *Figure 4-8*. Generally, pixel operation can be divided into two main stages, reset and phototransduction.

(a) The reset stage. During this stage, the photodiode capacitance is charged to a reset voltage by turning on the *Reset* transistor. This reset voltage is read out to one of sample-and-hold (S/H) in a correlated double sampling (CDS) circuit. The CDS circuit, usually located at the bottom of each column, subtracts the signal pixel value from the reset value. Its main purpose is to eliminate fixed pattern noise caused by random variations in the threshold voltage of the reset and pixel amplifier transistors, variations in the photodetector geometry and variations in the dark current. In addition, it should eliminate the $1/f$ noise in the circuit.

(b) The phototransduction stage. During this stage, the photodiode capacitor is discharged through a constant integration time at a rate

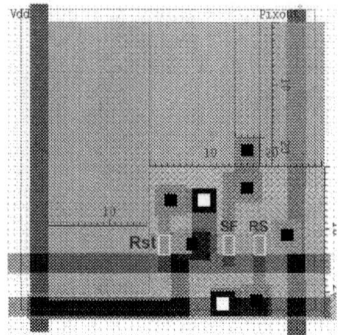

Figure 4-9. An example of a pixel layout with an L-shaped active area, which is the most common pixel design.

approximately proportional to the incident illumination. Therefore, a bright pixel produces a low analog signal voltage and a background pixel gives a high signal voltage. This voltage is read out to the second S/H of the CDS by enabling the row select transistor of the pixel. The CDS outputs the difference between the reset voltage level and the photovoltage level.

Because the readout of all pixels cannot be performed in parallel, a rolling readout technique is applied. All the pixels in each row are reset and read out in parallel, but the different rows are processed sequentially. *Figure 4-10* shows the time dependence of the rolling readout principle. A given row is accessed only once during the frame time (T_{frame}). The actual pixel operation sequence is in three steps: the accumulated signal value of the previous frame is read out, the pixel is reset, and the reset value is read out to the CDS. Thus, the CDS circuit actually subtracts the signal pixel value from the reset value of the next frame. Because CDS is not truly correlated without frame memory, the read noise is limited by the reset noise on the photodiode. After the signals and resets of all pixels in the row are read out to S/H, the outputs of all CDS circuits are sequentially read out using X-addressing circuitry, as shown in *Figure 4-7*.

The output photodiode signal is supposedly independent of detector size, because the lower pixel capacitance of smaller detectors causes an increase in conversion gain that compensates for the decrease in detector size. However, peripheral capacitances from the perimeters of the detector increase the total capacitance of the sensing node and thus decrease the conversion gain. As the pixel size scales down, photosensitivity decreases and the reset noise scales as $C^{1/2}$, where C is the photodiode capacitance. These tradeoffs must be considered when designing pixel fill factor, DR, SNR and conversion gain.

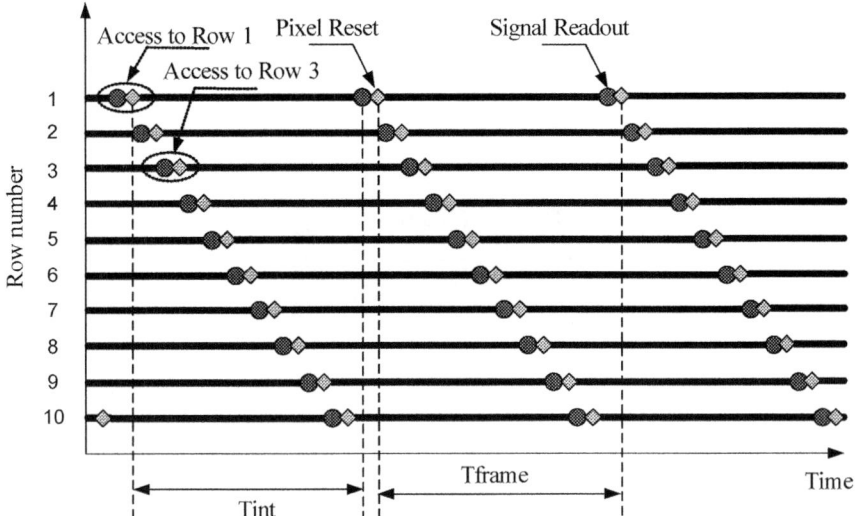

Figure 4-10. Rolling readout principle of the photodiode APS.

4.2.4   Photogate APS

Figure 4-11 shows the common photogate pixel architecture [10]. The basic concept for the photogate pixel arose from CCD technology. While photon-generated charge is integrated under a photogate with a high potential well, the output floating node is reset and the corresponding voltage is read out to the S/H of the CDS. When the integration is completed, the charge is transferred to the output floating node by pulsing the signal on the photogate. Then the corresponding voltage from the

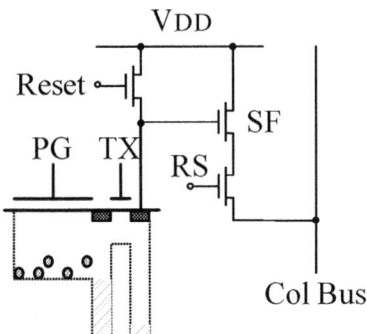

Figure 4-11. Basic photogate pixel architecture.

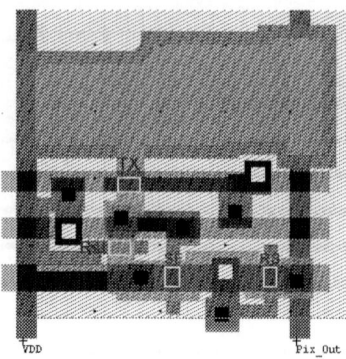

Figure 4-12. Example layout of a photogate pixel design.

integrated charge is read by the source follower to the second S/H of the CDS. The CDS outputs the difference between the reset voltage level and the photo-voltage level.

As mentioned above, the CDS can suppress reset noise, $1/f$ noise and FPN due to V_T and lithographic variations in the array. The reduction of noise level increases the total dynamic range and the SNR. The primary noise source for the photogate APS is photon shot noise, which cannot be suppressed by any means.

The photogate has a pixel pitch typically equal to 20 times the minimum size of the technology, since there are five transistors in each pixel. Due to the overlaying polysilicon, however, there is a reduction in QE, particularly in the blue region of the spectrum. The photogate pixel architecture is shown as a mask layout in *Figure 4-12*.

4.2.5   Pinned photodiode APS

The pinned photodiode pixel consists of a pinned diode (p$^+$-n-p), where the photon collection area is dragged away from the surface in order to reduce surface defect noise (such as that due to dark current) [13]. Photon-generated charge is integrated under a pinned diode and transferred to the output floating diffusion for the readout. As in the photogate APS, the sense node and integration node are separated to minimize noise. However, the primary difference is that the potential well for charge collection in a pinned diode is generated by a buried intrinsic layer (or an n-type layer) instead of a pulsed gate voltage as in the photogate. Each pinned diode pixel has four transistors and five control lines, resulting in a fill factor higher than in the photogate pixel, but lower than in the photodiode pixel. However, the small photon collection area of the pinned diode results in a very small full well

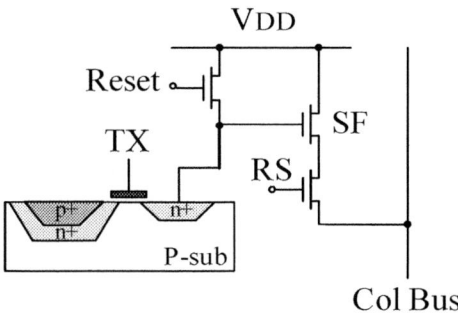

Figure 4-13. Basic pinned photodiode pixel architecture.

for photon-generated charge collection and lower QE compared to the photodiode pixel. *Figure 4-13* shows a basic pinned photodiode pixel structure.

4.2.6   Logarithmic Photodiode APS

Another type of APS is a logarithmic photodiode sensor [11]. This three-transistor pixel enables logarithmic encoding of the photocurrent, thus increasing the dynamic range of the sensor; i.e., the same range of sensor output voltage is suitable for a wider range of illumination. The implementation of the basic logarithmic photodiode pixel is described in *Figure 4-14*.

This pixel does not require reset and operates continuously. The voltage on the photodiode (V_{PH}) is approximately equal to V_{DD}, causing the load transistor to operate in the subthreshold region ($V_{DD} = V_{PH} + \Delta V_{PH}$). The

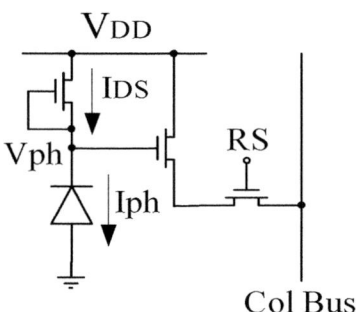

Figure 4-14. Basic logarithmic photodiode pixel architecture.

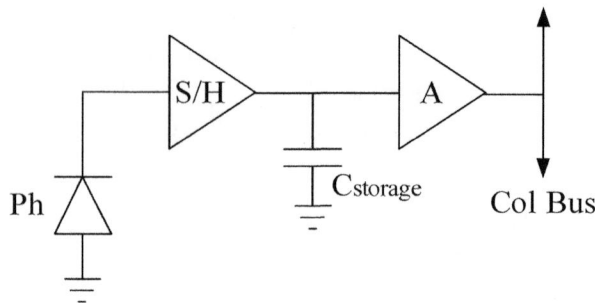

Figure 4-15. Architecture of a snapshot photodiode APS (after [12]).

photocurrent (I_{PH}) is equal to the subthreshold current (I_{DS}). The voltage on the photodiode decreases logarithmically with linear increases in illumination (and thus photocurrent) following the equation

$$V_{PH} = V_{DD} - \Delta V_{PH} = V_{DD} - \frac{KT}{q} \cdot \ln\left(\frac{I_{PH}}{I_0}\right) \qquad (4.1)$$

where KT/q is 0.026 V at $T = 300$ K and I_0 represents all constant terms. While the logarithmic APS has the above-mentioned advantages, it suffers from serious drawbacks such as significant temperature dependence of the output, low swing of the output (especially for relatively low illumination levels) and high FPN. Accordingly, the logarithmic APS is not commonly used.

4.2.7   Snapshot pixels

Most CMOS imagers feature the rolling shutter readout method (also known as the rolling readout method) shown in *Figure 4-10*. In the rolling shutter approach, the start and end of the light collection for each row is slightly delayed from the previous row; this leads to image deformity when there is relative motion between the imager and the scene. The ideal solution for imaging objects moving at high speed is the snapshot imager, which employs the electronic global shutter method [12]. This technique uses a memory element inside each pixel and provides capabilities similar to a mechanical shutter: it allows simultaneous integration of the entire pixel array and then stops the exposure while the image data is read out. The principal scheme of snapshot photodiode APS was introduced by Yadid-Pecht in 1991, and is shown in *Figure 4-15*.

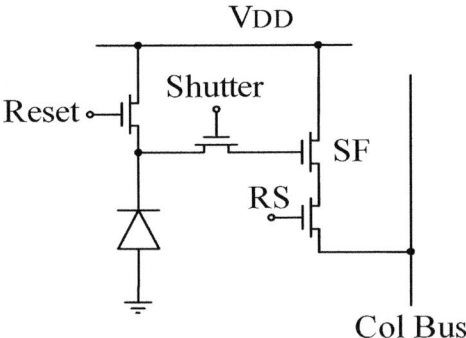

Figure 4-16. Transistor scheme of a commonly used snapshot pixel.

The snapshot pixel includes a sample-and-hold (S/H) switch with analog storage, which consists of all parasitic capacitances in the amplifier input. The in-pixel amplification is performed by a source follower amplifier, identical to that in a rolling shutter pixel. The full transistor scheme of a commonly used global shutter pixel is shown in *Figure 4-16*.

In contrast to the rolling shutter technique, a sensor with global shutter architecture exposes all its pixels at the same time. After the integration time T_{int}, the signal charge is stored in an in-pixel sample-and-hold capacitance until readout. One of the problems that should be addressed in the snapshot pixels is the shutter efficiency. The light exposure of the S/H stage, shutter leakage and the limited storage capacitance lead to signal lost. *Figure 4-16* shows a pixel that employs an NMOS transistor as a shutter. This implementation allows a small pixel area, but it has a low shutter efficiency. Shutter efficiency can be increased using a PMOS transistor as a shutter, if it is well separated from the photodiode. Unfortunately, a PMOS shutter decreases the fill factor and, due to increased parasitic capacitances, also decreases the conversion gain.

4.3 APS system-on-a-chip approach

CMOS image sensors allow implementations of complex sensing systems on a single silicon die. For example, almost all CMOS imagers employ analog to digital conversion on the same die. There are three general approaches to implementing ADC with active pixel sensors:

1. *Chip-level ADC*, where a single ADC circuit serves the whole APS array. This method requires a very high-speed ADC, especially if a very large array is used.
2. *Column-level ADC*, where an array of ADCs is placed at the bottom of the APS array and each ADC is dedicated to one or more columns

of the APS array. All these ADCs are operated in parallel, so a low-to-medium-speed ADC design can be used, depending on the APS array size. The disadvantages of this approach are the necessity of fitting each ADC within the pixel pitch (i.e., the column width) and the possible problems of mismatch among the converters on different columns.

3. *Pixel-level ADC*, where every pixel has its own converter. This approach allows parallel operation of all ADCs in the APS array, so a very low speed ADC is suitable. Using one ADC per pixel has additional advantages, such as higher SNR, lower power and simpler design.

In this section, two examples of CMOS imagers are described. The first implements a "smart vision" system-on-a-chip based on a photodiode APS with linear output over a wide dynamic range. An increase in the dynamic range of the sensors is enabled by random access to the pixel array and by the insertion of additional circuitry within each pixel. The second example is a smart tracking sensor employing analog nonlinear winner-take-all selection.

In the design of a "smart" sensing system, an important step is to decide whether computation circuitry should be inserted within the pixel or placed in the periphery of the array. When processing circuitry is put within the pixel, additional functions can be implemented, simple 2-D processing is possible, and neighboring pixels can be easily shared in neural networks. These systems are also very useful for real-time applications. On the other hand, the fill factor is drastically reduced, making these systems unsuitable for applications where high spatial resolution and very high image quality are required. In all systems presented later in this chapter, most of the processing circuitry is placed in the periphery of the array to avoid degradation of image quality.

4.3.1   Autoscaling APS with customized increase of dynamic range

This section introduces the reader to the dynamic range problem in CMOS imagers, showing possible existing solutions. Then an advanced autoscaling CMOS APS with customized linear increase of DR is explained.

4.3.1.1   Dynamic range problem and possible solutions

Scenes imaged with electronic cameras can have a wide range of illumination. Levels can range from 10^{-3} lux for night vision to 10^5 lux for scenes illuminated with bright sunlight, and even higher levels can occur with the direct viewing of light sources such as oncoming headlights. The intrascene dynamic range capability of a sensor is measured as

$$DR = 20 \cdot \log(S/N) \qquad (4.2)$$

where S is the saturation level and N is the root mean square (rms) read noise floor measured in electrons or volts. The human eye has a dynamic range of about 90 dB and camera film of about 80 dB, but typical CCDs and CMOS APS have a dynamic range of only 65–75 dB. Generally, dynamic range can be increased in two ways: the first one is noise reduction and thus expanding the dynamic range toward darker scenes. The second method is incident light saturation level expansion, thus improving the dynamic range toward brighter scenes.

Bright scenes and wide variations in intrascene illumination can arise in many situations: driving at night, photographing people in front of a window, observing an aircraft landing at night, and imaging objects for studies in meteorology or astronomy. Various solutions have been proposed in both CCD and CMOS technologies to cope with this problem [13–37]. Methods for widening the dynamic range can be grouped into five areas:

1. *Companding sensors,* such as logarithmic compressed-response photodetectors;
2. *Multi-mode sensors,* where operation modes are changed;
3. *Frequency-based sensors,* where the sensor output is converted to pulse frequency;
4. *Sensors with external control over integration time,* which can be further divided into global control (where the integration time of the whole sensor can be controlled) and local control (where different areas within the sensor can have different exposure times); and
5. *Sensors with autonomous control over integration time,* in which the sensor itself provides the means for different integration times.

Companding sensors. The simplest solution to increase DR is to compress the response curve using a logarithmic photodiode sensor, as described in section 4.2.6. Another type of companding sensor for widening the DR was introduced by Mead [16], where a parasitic vertical bipolar transistor with a logarithmic response was used. The logarithmic function there is again a result of the subthreshold operation of the diode-connected MOS transistors, added in series to the bipolar transistor. The voltage output of this photoreceptor is logarithmic over four or five orders of magnitude of incoming light intensity. It has been used successfully by Mahowald and Boahen [17]. This detector operates in the subthreshold region and has a low output voltage swing. A disadvantage of these pixels is that this form of

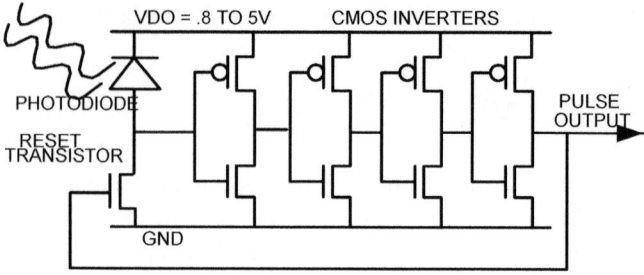

Figure 4-17. A pulse photosensor with reset circuitry (after Yang).

compression leads to low contrast and loss of details; adaptation, where linearity around the operation point is exploited, was proposed to alleviate this problem [19–20]. In addition, the response of the logarithmic pixels with this kind of readout is light dependent. This means that for low light intensities the readout time would be very slow, depending also on the photodiode capacitance.

Multimode sensors. A multisensitivity photodetector was proposed by Ward et al. [21]. The detector is a parasitic vertical bipolar transistor between diffusion, well and substrate. By connecting a MOS transistor to the base and the emitter of the bipolar transistor in a Darlington structure, the current gain can be boosted further. Thus, both bipolar transistors can be activated at very low light intensities and inactivated at higher intensities. For moderate levels, only one bipolar transistor is activated. Two selection transistors are required within the pixel for choosing the mode. This pixel occupies a relatively large area.

Frequency-based sensors. In 1994, Yang [22] proposed a pulse photosensor that uses simple integrate-and-reset circuitry to directly convert optical energy into a pulse frequency output. The output of this photosensor can vary over 5–6 orders of magnitude and is linearly proportional to optical energy. A diagram of the circuitry is presented in *Figure 4-17*.

The pixel fill factor is much decreased with this approach, since the inverter chain resides next to the photodiode. In addition, the pulse timing relies on the threshold voltages of the inverters. Since threshold voltage mismatches exists between different transistors, there will be a different response for each pixel. This makes this sensor worse in terms of noise, since the threshold mismatch translates to a multiplicative error (the output frequency of the pulses is affected) and not just constant FPN.

Sensors with external control over integration time. A multiple-integration-time photoreceptor has been developed at Carnegie Mellon University [23]. It has multiple integration periods, which are chosen depending upon light intensity to avoid saturation. When the charge level nears saturation, the integration is stopped at one of these integration periods and the integration time is recorded. This sensor has a very low fill factor.

Sensors with autonomous control over integration time. The automatic wide-dynamic-range sensor was proposed by Yadid-Pecht [24, 25]. This imager consisted of a two-dimensional array of sensors, with each sensor capable of being exposed for a different length of time with autonomous on-chip control. Reset enable pulses are generated at specific times during the integration period. At each reset enable point, a nondestructive readout is performed on the sensor and compared to a threshold value. A conditional reset pulse is generated if the sensor value exceeds the threshold voltage. The main drawback with this solution is the extent of the additional circuitry, which affects spatial resolution.

In the following sections, we describe an APS with an in-pixel autoexposure and a wide-dynamic-range linear output. Only a minimum additional area above the basic APS transistors is required within the pixel, and the dynamic range enhancement is achieved with minimal effects on the temporal resolution.

4.3.1.2 System architecture

The architecture of the DR approach is shown in *Figure 4-18*. As in a traditional rolling-shutter APS, this imager is constructed of a two-dimensional pixel array, here of 64 columns and 64 rows, with random pixel access capability. Each pixel contains an optical sensor to receive light, a reset input and an electrical output representing the illumination received. The pixel used here is not a classic pixel, since it enables individual pixel reset via an additional transistor [5]. The outputs of a selected row are read through the column-parallel signal chain, and at certain points in time are also compared with an appropriate threshold in the comparison circuits. If a pixel value exceeds the threshold, a reset is given at that time to that pixel. The binary information concerning the reset (i.e., applied or not) is saved in digital storage for the later calculation of the scaling factor. The pixel value can then be determined as a floating-point number, where the exponent comes from the scaling factor for the actual integration time and the mantissa from the regular A/D output. Therefore, the actual pixel value would be

$$Value = Man \cdot \left(T / \left(T / X^{EXP} \right) \right) = Man \cdot X^{EXP} \qquad (4.3)$$

where *Value* is the actual pixel value, *Man* (mantissa) is the analog or

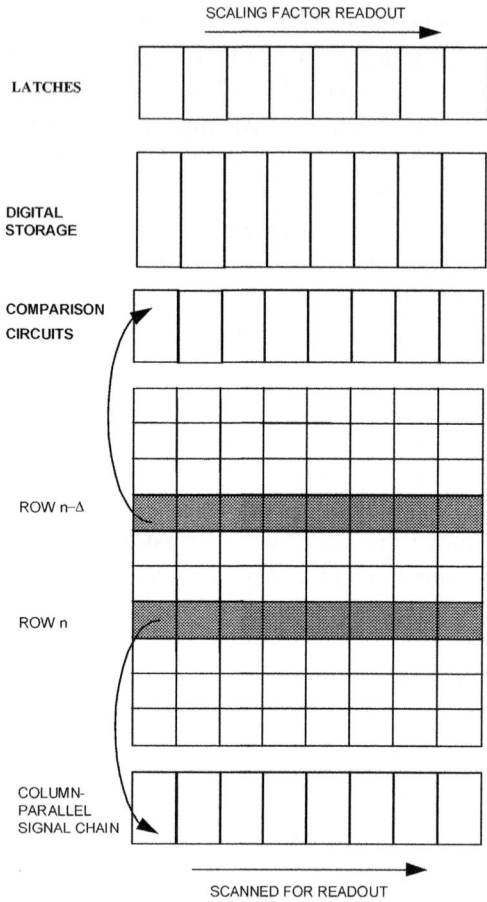

Figure 4-18. General architecture description of the autoscaling CMOS APS.

digitized output value that has been read out at the time T, X is a constant greater than one (for example, 2), and *EXP* is the exponent value describing the scaling factor. This digital value is read out at the upper part of the chip. For each pixel, only the last readouts of a certain number of rows are kept to enable the correct output for the exponent bits.

The idea of having a floating-point presentation per pixel via real-time feedback from the pixel has been proposed before by Yadid-Pecht [36]. However, the previous design required an area in the pixel that substantially affected the fill factor, so it was then proposed that the control for a group of pixels should be shared. In the currently proposed solution, however, the necessary additional hardware will be placed in the periphery; as a result, the

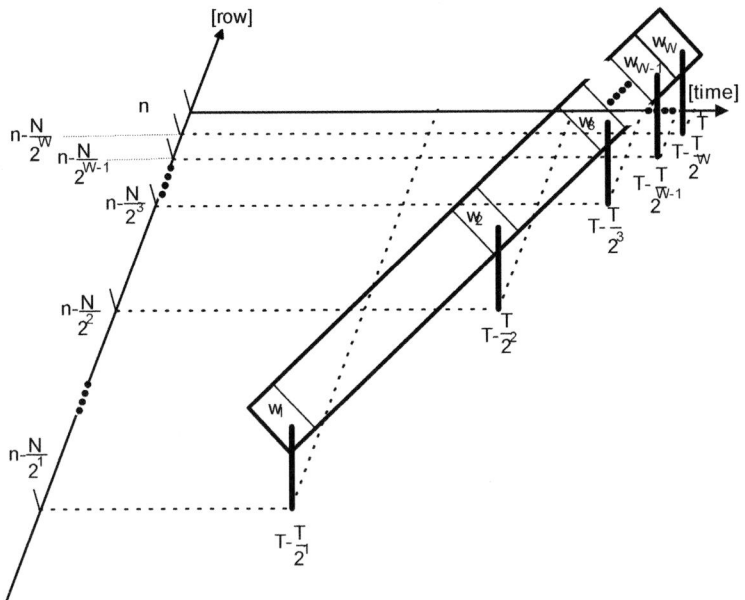

Figure 4-19. Combined time-space diagram.

information can be output with minimal effect on spatial or temporal resolution. The spatial resolution is slightly modified, since the desired ability to independently reset each pixel requires an additional transistor per pixel [37]. The temporal resolution should be assessed: a pixel should be checked at different time points to get the exponential (scaling) term. Equivalently, the pixels of different rows could be checked to get the same information. In the latter case, row n at time zero would provide the mantissa for row n (either through an on-chip or an off-chip A/D output), while the pixels in row $n - N/2$ (where N is the total number of rows that set the frame time) would provide the first exponent bit (W_1) as a result of the logic circuit decision for that row. Row $n - N/4$ would provide the second bit (W_2) for that row, row $n - N/8$ would provide the third bit, and so on. Thus, at the cost of a customized number of comparisons, the required information can be obtained automatically and the mantissa scaled accordingly.

Figure 4-19 describes this approach via a combined time-space diagram where the axes represent the row number and time, respectively. W_1, W_2, ... W_w, represent the exponent bits; i.e., W_1 represents the first point of decision $T - T/2$ (whether to reset or not for the first time), W_2 for the next point and so forth. The equivalent is shown at point $n - N/2$ in the spatial domain, which is the row that should be used for the decision concerning W_1.

For example, an imaging device consisting of a square pixel array of 64 rows and 64 columns may be assumed, and it is desired to expand the dynamic range by 3 bits due to a high illumination level. Therefore, $W = 3$ and $N = 64$ in this case. For each pixel from the selected row n, three comparisons (W_1, W_2 and W_3) are carried out at three corresponding time points ($T - T/2$, $T - T/4$, $T - T/8$). The first comparison, which is with the threshold, is carried out at row $n - 32$ (32 rows before regular readout of that pixel). This leaves an integration time of $T/2$ with a comparison result of $W_1 = $ "1", and this pixel is reset. The second comparison is carried out at row $n - 16$ (16 rows before regular readout of that pixel), leaving an integration time of $T/4$ with a comparison result of $W_2 = $ "1"; this pixel is also reset. The third comparison is carried out at row $n - 8$ (8 rows before regular readout of that pixel), leaving an integration time of $T/8$ with a comparison result of $W_3 = $ "1". This pixel is reset as well. The autoscaling combination for the pixel in this example is therefore (1 1 1): this means that the pixel has been reset three times during the frame time, and the regular readout for this pixel should be scaled (multiplied) by a factor of $2^3 = 8$.

The frame time of a pixel array consisting of N rows and N columns may be calculated. Readout time for each row is composed of a copying time ($T_{copy} = $ the time needed to copy one row into the readout buffer) and a scanning time ($T_{scan} = $ the time needed to scan each pixel). Since there are N pixels in each row, the total time for row readout (T_{row}) is given by

$$T_{row} = T_{copy} + N \times T_{scan} \tag{4.4}$$

and the frame time (T_{frame}) is given by

$$T_{frame} = N \times T_{scan} \tag{4.5}$$

By adding W comparisons (for W different integration times) for each row, the row readout time is slightly modified and is given by

$$\begin{aligned} T'_{row} &= W \times T_{comp} + T_{row} \\ &= W \times T_{comp} + T_{copy} + N \times T_{scan} \end{aligned} \tag{4.6}$$

where T_{comp} is the time for comparison to the threshold level. Since $W \ll N$ then

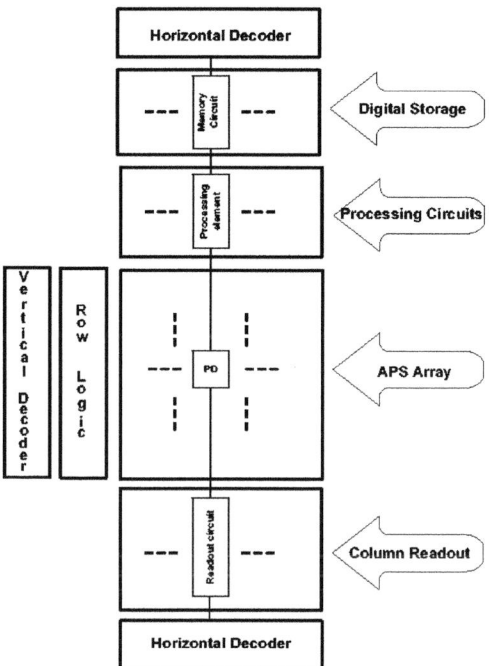

Figure 4-20. Block diagram of the proposed design.

$$W \times T_{comp} \ll N \times T_{scan} \qquad (4.7)$$

and $T'_{row} = T_{row}$. Hence, the frame time T_{frame} is insignificantly affected by autoscaling. This enables the imager to process scenes without degradation in the frame rate.

4.3.1.3 Design and implementation

A block diagram of the proposed design is shown in *Figure 4-20*. The design makes use of a column parallel architecture to share the processing circuits among the pixels in a column. The pixel array, the memory array and the processing elements are separated in this architecture. Each pixel contains an additional transistor (in series with the row reset transistor) that is activated by a vertical column reset signal; this allows the independent reset of the pixel. Integration time can be adjusted for each pixel with this reset, and nondestructive readout of the pixel can be performed at any time

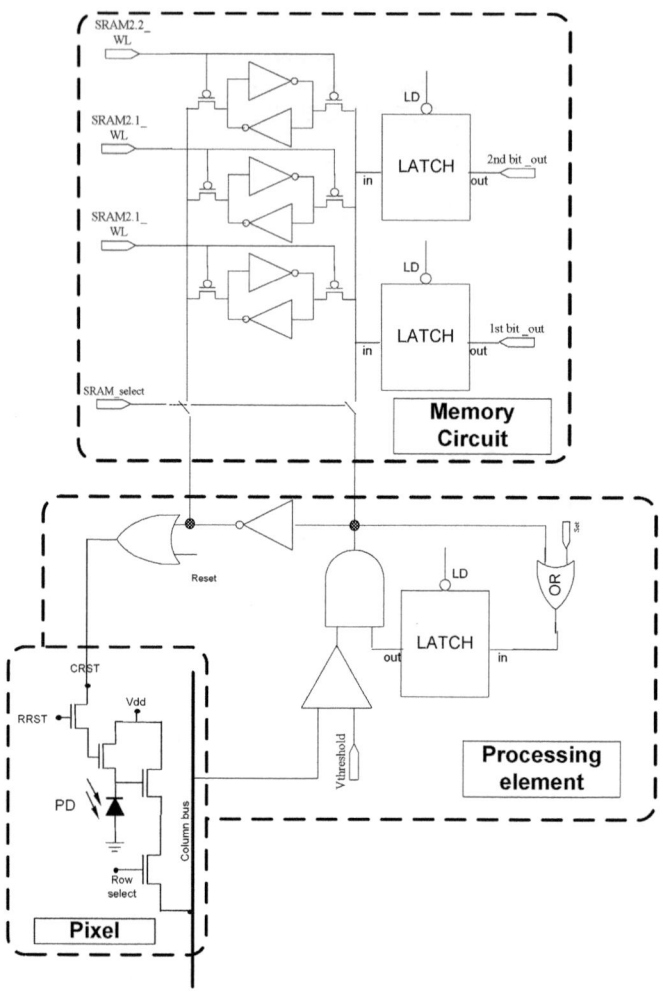

Figure 4-21. Electrical description of one column of an autoscaling CMOS APS.

during the integration period by activating the row select transistor and reading the voltage on the column bus.

The processing element contains the saturation detection circuit, which is shared by all pixels in a column. Because of the column parallel architecture, the pixel array contains a minimum amount of additional circuitry and sacrifices little in fill factor. The memory array contains the SRAMs and latches. Two horizontal decoders—one each for the pixel array and the memory array—work in parallel and are used to retrieve the mantissa and exponent, respectively. The vertical decoder is used to select the rows in order.

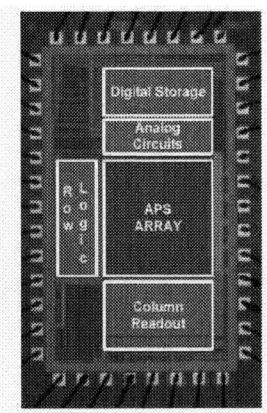

Figure 4-22. Photograph of the fabricated autoscaling CMOS APS test chip.

The electrical scheme for a single column is presented in *Figure 4-21*. In this circuit, the pixel output signal is evaluated at the comparator, where it is compared with an appropriate threshold. If its signal exceeds a predetermined threshold, the pixel is detected as saturated. Using this information and the binary information concerning the pixel (stored in the memory during different parts of the integration), a decision whether to reset the pixel is made. If the decision is positive, the column reset (*CRST*) and row reset (*RRST*) lines must both be precharged at a logical high voltage to activate the reset transistor; the photodiode then restarts integration. If the decision is negative, the reset is not active and the pixel continues to integrate. The binary information (whether the reset was applied or not) is saved in the SRAM memory storage and output to the latches in due time. After the row is read through the regular output chain, this additional information is retrieved from the memory through the latches.

4.3.1.4 Experimental results

A 64×64 pixel chip was successfully fabricated using the HP 0.5 µm n-well process. The chip photograph is shown in *Figure 4-22*. The sensor was quantitatively tested for relative responsivity, conversion gain, saturation level, noise, dynamic range, dark current, and fixed pattern noise. The results are presented in *Table 4-1*.

The conversion gain was in general agreement with the design estimate of photodiode capacitance. The saturation level was approximately 1.33 V; fixed pattern noise (FPN) was measured to be approximately 0.15% saturation; dark current was measured to be on the order of 30–35 mV/sec, output referred, or 0.61 pA/cm^2; and the inherent dynamic range was

Table 4-1. Attributes of the autoscaling CMOS APS test chip.

Chip format	64 × 64 pixels
Chip technology	HP 0.5 μm
Chip size	1.878 mm × 2.9073 mm
Pixel size	14.4 μm × 14.4 μm
Pixel type	Photodiode
Pixel fill factor	37%
Conversion gain	12 μV/e⁻
Fixed pattern noise (FPN)	0.15%
Dark current (room temp)	35 mV/sec (0.61 pA/cm^2)
Power	3.71 mW (5 MHz)
Inherent dynamic range	71.4 dB (~11 bit)
Extended dynamic range	2 additional bits
Saturation level	1.33 V
Quantum efficiency (QE)	20%

71.4 dB, or 11 bits. The extended dynamic range provided two additional bits to the inherent dynamic range. No smear or blooming was observed due to the lateral overflow drain inherent in the APS design. The chip was also functionally tested.

Figure 4-23 shows a comparison between an image captured by a traditional CMOS APS and by the autoexposure system presented here. In the *Figure 4-23(a)*, a scene is imaged with a strong light on the object; hence, some of the pixels are saturated. At the bottom of *Figure 4-23(b)*, the capability of the autoexposure sensor for imaging the details of the saturated area in real time may be observed. Since the display device is limited to eight bits, only the most relevant eight-bit part (i.e., the mantissa) of the thirteen-bit range of each pixel is displayed here. The exponent value, which is different for different areas, is not displayed. This concept in its present form suits rolling-shutter sensors, and a first prototype following this concept has been demonstrated here.

4.3.2   CMOS smart tracking sensor employing WTA selection

This section presents an example of a smart APS sensor suitable for tracking purposes. The system employs an analog winner-take-all circuit to find and track the brightest object in the field of view (FOV). This system-on-a-chip employs adaptive spatial filtering of the processed image, with elimination of bad pixels and with reduction of false alarm when the object is missing. The circuit has a unique adaptive spatial filtering ability that allows the removal of the background from the image, and this occurs one stage before the image is transferred to the WTA detection circuit. A test chip of 64 × 64 array has been implemented in 0.5 μm CMOS technology. It

Figure 4-23. (a) Scene observed with a traditional CMOS APS sensor.(b) Scene observed with the in-pixel autoexposure CMOS APS sensor.

has a 49% fill factor, it is operated by 3.3 V supply, and it dissipates 36 mW at video rate. The system architecture and operation are described, together with measurements from a prototype chip.

4.3.2.1 Motivation

Many scientific, commercial and consumer applications require spatial acquisition and tracking of the brightest object of interest. A winner-take-all function has an important role in these kinds of systems: it selects and identifies the highest input (which corresponds to the brightest pixel of the sensor) and inhibits the rest. The result is a high digital value assigned to the winner pixel and a low one assigned to the others. CMOS implementations of WTA networks are an important class of circuits widely used in neural networks and pattern-recognition systems [42, 44]. Many WTA circuit implementations have been proposed in the literature [40–52]. A current-mode MOS implementation of the WTA function was first introduced by Lazzaro [40]. This very compact circuit optimizes power consumption and silicon area usage. It is asynchronous, responds in real time and processes all input currents in parallel.

Most of the existing WTA circuits can be integrated with APS sensors. Usually, when WTA circuits are used in two-dimensional tracking and visual attention systems, the image processing circuitry is included in the pixel; however, there are penalties in fill factor or pixel size and resolution. Here we show an implementation of a 2-D tracking system using 1-D WTA circuits. All image processing is performed in periphery of the array without influencing imager quality.

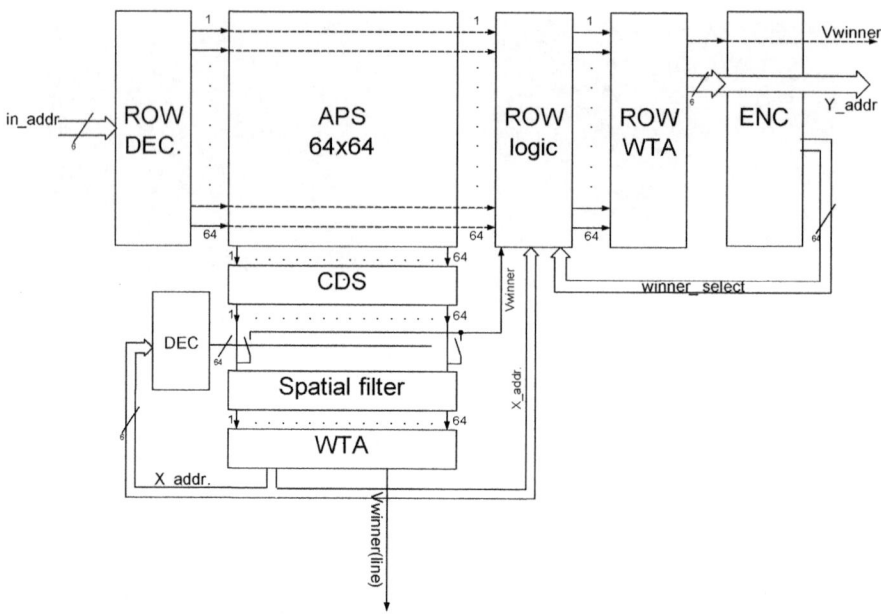

Figure 4-24. Block diagram of the proposed CMOS smart tracking sensor system.

As mentioned above, the regular WTA circuit chooses a winner from a group of input signals. When an APS with WTA selection system is used for object selection, a number of problems can occur. If the object of interest disappears from the processed image, the WTA will compare input voltages that represent intensity levels in the image background; hence, the circuit will output the coordinates of some background pixel instead of the missing object of interest and cause a false alarm. Background filtering and false alarm reduction is therefore necessary. Another problem that can disrupt proper operation of the system is a bad pixel. Since the bad pixel has a high value, it can be selected as the winner regardless of other pixel values.

The simplest technique for background filtering is signal comparison against a globally set threshold, above which pixels qualify as object pixels. The disadvantage of this kind of filtering is that it is necessary to choose the value of this threshold in advance; in the case where the background is overly bright (i.e., above the chosen threshold), the circuit will not be able to cope with the task. The problem is most severe if the object of interest disappears from the processed image and the background is bright.

The system described here is a 64 × 64 element APS array with two-dimensional WTA selection. A spatial adaptive filtering circuit allows adaptive background filtering and false alarm reduction if the object is missing from the scene.

4.3.2.2 System architecture

Figure 4-24 shows the block diagram of the system. There is no penalty in spatial resolution using the proposed architecture, since the processing is done at the periphery of the array.

The process of phototransduction in the APS is done row-by-row in two phases: reset (where a photodiode capacitor is charged to a high reset voltage) and phototransduction (where the capacitor is discharged after a constant integration time). A commonly used CDS circuit is implemented on the chip to subtract the signal pixel value from the reset one; the output is high for a bright pixel and low for a dark one. The subsequent stage is adaptive background filtering of all pixels for which the CDS values are less than a threshold value. This threshold corresponds to the average of the outputs from all row sensors, with the addition of a small variable epsilon value. Only the signals that pass the filter are transmitted to the WTA selection circuit. When there is no detectable object (i.e., only background signals exist), no signals pass this filter, and so the "no object" output becomes high and false alarm reduction is achieved. A detailed description of this filtering technique is presented in section 4.3.2.3.

The next stage is the winner-take-all selection, which is done with a simple voltage-mode WTA after Donckers et al. [48]. The main factor for choosing this WTA circuit is its simplicity; generally any kind of voltage- or current-mode WTA can be integrated with this system [53]. The winner selection is done row-by-row. The winner of each row is found, its value is stored into an analog memory (a capacitor), and its address is deposited in the digital storage. If there is more than one input with the same high value, the WTA chooses the leftmost one using a simple logic. The result is a column of N analog pixel values of row winners with their column addresses. From all the row winners, a global winner is selected using an identical WTA circuit. The 1-D winner selection array was designed to consist of eight identical blocks of eight-input WTA cells to achieve better resolution and reduce matching problems. The row winner value is stored in the analog memory that corresponds to the actual row and its column address is stored in the corresponding digital memory; these analog and digital memories are in the *ROW logic* block displayed in *Figure 4-24*. In the case of "no object" in a row, the value transmitted to the memory is zero. Following a full frame scan, the WTA function is activated on all 64 row winners (in the *ROW WTA* block in *Figure 4-24*) and the location of the global frame winner is found. Its analog value and its address are then read out of the memory by an encoder (the *ENC* block in *Figure 4-24*).

This architecture allows the enlargement of the proposed system to any size of $N \times N$ pixel array without affecting the system properties.

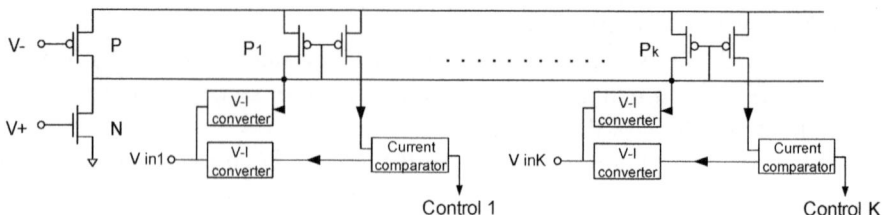

Figure 4-25. Principal scheme of the spatial filter circuit in the CMOS smart tracking sensor.

4.3.2.3   Descriptions of the circuits

4.3.2.3.1 The adaptive spatial filter

The principal scheme of the adaptive 1-D filter circuit used in the CMOS smart tracking sensor system is shown in *Figure 4-25*. The inputs to the filter correspond to the CDS values, and the outputs are the control signals. As mentioned earlier, this circuit filters all pixels for which the CDS values are less than a threshold value. This threshold corresponds to the average of the outputs from all the row sensors, with the addition of a small variable epsilon value. The *Control* output of the filter is high for an object pixel and low for a pixel that is imaging the background. The advantage of this filtering is that this epsilon value is adaptive and not constant for different input vectors. The value of epsilon depends inversely on the average current value: it increases when the average current decreases and decreases when current increases. This results in a small epsilon value for a high background and a high epsilon for a low background. The filtering process is thus more sensitive when a high background is present and the input voltage of the object is very similar to that of the background. The epsilon value can be controlled manually by setting suitable *V*– and *V*+ voltage values (see *Figure 4-25*).

If *V*+ is zero and *V*– is V_{DD}, then the epsilon value is zero and all $I_{average} + \varepsilon$ currents of the circuit are equal to the average current of the *n* inputs in an *n*-sized array. The non-zero epsilon value can be added to this average by increasing the *V*+ value. The epsilon value can also be subtracted from the average by decreasing the *V*+ value. A positive epsilon value is usually of interest here, so *V*– is V_{DD} in this case. Note that the voltages to the current converters have a pull configuration (the current direction is shown by arrows in *Figure 4-25*).

The adaptive functionality can be achieved by operation of transistor N in the linear region. With an average current increase (reflecting an increase in background), the V_{gs} values of the $P_1 \ldots P_k$ transistors are increased as well,

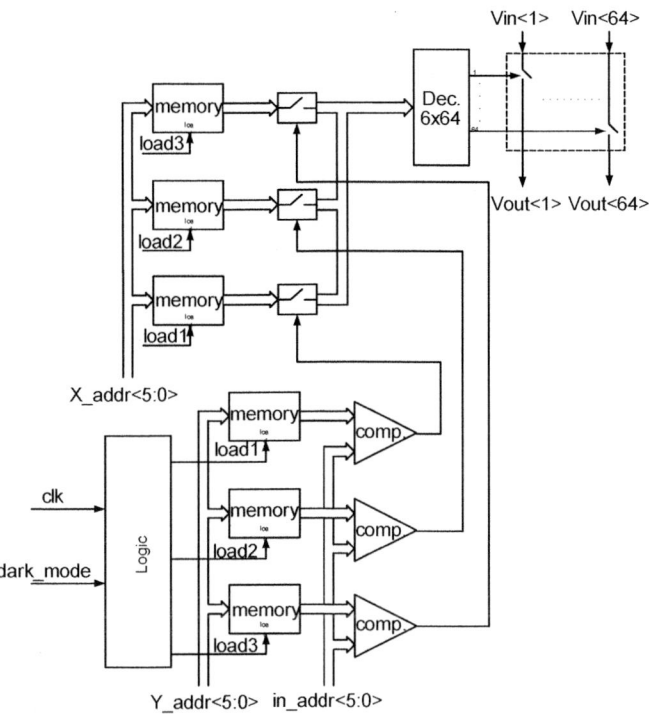

Figure 4-26. Principal scheme for bad pixel elimination in the CMOS smart tracking sensor.

which causes a reduction in the V_{ds} voltage of transistor N. The result is a fall in current at transistor N and a reduction of epsilon. Note that if a constant epsilon value is required, a stable current source (independent of the V_{sg} of $P_1...P_k$) can be used instead of transistor N.

Instead of using a transistor N, another simple option is to connect two parallel transistors, one with a relatively large W/L value operating in saturation and another with a small W/L operating in the linear region. This arrangement can achieve a constant epsilon by cutting off the transistor that is usually operated in the linear region and using only the saturation transistor. Alternatively, an adaptive epsilon value can be achieved by using the transistor operating in the linear region and cutting off the transistor that is in saturation.

In addition to background filtering, the circuit enables "no object" notification. If the control values of all pixels are "0", the "no object" output is high and false alarms are reduced. However, noise levels vary with the background level and shot noise is higher at a higher background level. To achieve proper operation of the "no object" function, the minimum value of epsilon must therefore be limited.

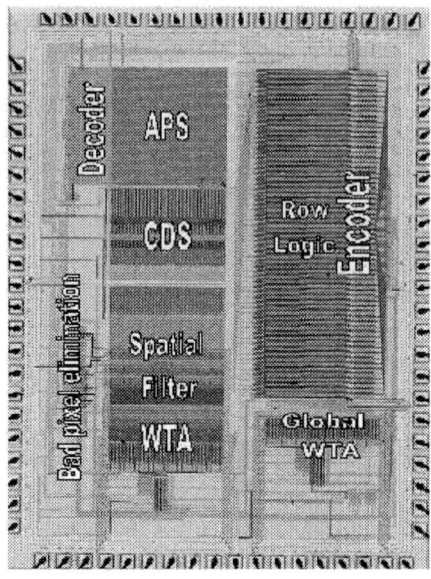

Figure 4-27. Photograph of the CMOS smart tracking sensor test chip.

The inputs to the WTA circuit depend on the filter control values: zero for a background pixel and pixel intensity level for an object pixel.

4.3.2.3.2 Elimination of bad pixels

Bad pixels can disrupt proper operation of the system. A bad pixel can have a high value, so it may be selected as the winner regardless of other pixel values. In the proposed CMOS smart tracking sensor system, bad pixels are disabled with a special "dark mode" in which a dark image is input. *Figure 4-26* shows the principal scheme for bad pixel elimination.

Bad pixels are eliminated in two stages. In the first stage (the dark mode), the *dark_mode* signal in *Figure 4-26* is "1" and the system processes a dark image—a very low background without an object of interest. The circuit finds bad bright pixels, using a regular WTA algorithm as described before. The frame must be scanned (processed) N times in order to find N bad pixels. After each additional frame scanning, a new bad pixel is found and its address is stored in the memory (using the X_addr and Y_addr buses in *Figure 4-26*).

In the second stage (regular system operation), the *dark_mode* signal is "0" and a real image is processed. For the "bad" pixels that were found in the dark mode stage, however, only ground values are transmitted to the filter and the WTA circuits. This is accomplished by comparing the address of every processed row (*in_addr* in *Figure 4-26*) with the Y_addr stored in

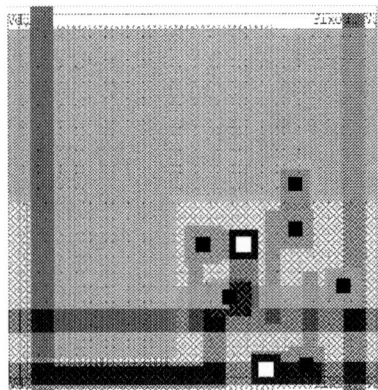

Figure 4-28. Layout of a single photodiode pixel of the CMOS smart tracking sensor.

the memory. If their values are equal, the comparator will output "1" and the *X_addr* of the bad pixel of this row is transferred to the *Dec. 6×64* block. In the array, the input bad pixel voltage is replaced with a ground value. The values of all bad pixels are replaced by ground, and thus bad pixels cannot be chosen as winners.

The fabricated prototype chip allows the elimination of up to three bad pixels. Adding more logic can easily increase this capacity.

4.3.2.4   Performance and test chip measurements

The CMOS smart tracking sensor system was designed and fabricated in a 0.5 μm, n-well, 3-metal, CMOS, HP technology process supported by MOSIS. The supply voltage was 3.3 V. A photograph of the test chip is shown in *Figure 4-27*. The test chip includes an APS array, row decoders, correlated double sampling circuit, an adaptive spatial filter, eight identical eight-input voltage WTA circuits, logic, a global winner selection circuit and a bad pixel elimination block.

The test chip was designed to allow separate modular testing of every functional block of the chip as well as measurements of the chip as a unit. The main separate blocks of the chip (the APS with CDS, the adaptive filter and the eight-input WTA circuit) were tested, as was the whole chip to check the functionality.

4.3.2.4.1 APS with CDS

The layout of a single photodiode pixel of the CMOS smart tracking sensor is shown in *Figure 4-28*.

As mentioned before, there is no penalty in spatial resolution for this architecture since the processing is done at the periphery. The pixel size is

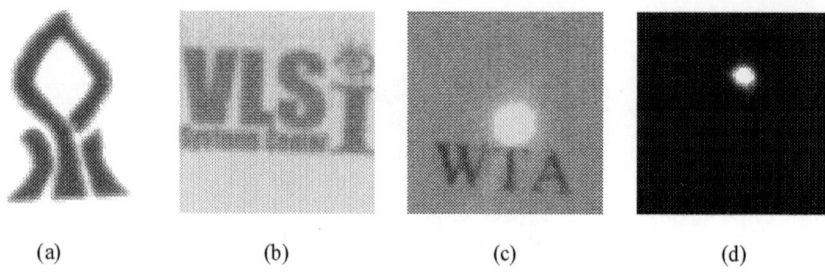

(a)   (b)   (c)   (d)

Figure 4-29. Four images as captured by the APS: (a) the Ben-Gurion University logo on a high illumination background; (b) the VLSI Systems Center logo; (c) WTA with a laser beam spot; and (d) a laser beam spot on a dark background.

14.4 μm × 14.4 μm and the fill factor is 49%. *Figure 4-29* shows four images as captured by the sensor under different background illumination levels.

4.3.2.4.2 Adaptive filter measurements

Measurements were carried out at both low and high background levels to determine the properties of the filter. These measurements check the ability of the circuit to filter background regardless of its level and also check the dependence of the ε value on the background level. *Figure 4-30(a)* and *(b)* show the response of the filter to low and high backgrounds, respectively.

In the first test, the average of the input voltages for a low background was 820 mV; this is represented by the horizontal solid line in *Figure 4-30(a)*. One of the filter inputs ranged from 0 V to 3.3 V; these inputs corresponded to the background in the case of a low value and the object in the case of a high value. They are represented by the sloped sawtooth lines in *Figure 4-30*. The pulsed voltage in *Figure 4-30* is the filter control output (see *Figure 4-25*). This square wave is low for $V_{in3} < 1.6$ V and high for $V_{in3} > 1.6$ V. This value represents $V_{average} + \varepsilon$ when $V_{in3} = 1.6$ V. In this case, the ε value is 780 mV. As mentioned earlier, the ε value can be changed for this input vector by changing the $V+$ and $V-$ control voltages.

The same procedure was performed to test the high background case, where the average was 1.54 V and the ε value was found to be 360 mV.

The filter responded as expected: a high epsilon value was generated for a low background level and a low epsilon value for a high background.

Figure 4-31 plots the epsilon value as function of background levels for different $V+$ values. As expected, an increase in $V+$ bias at a constant background level gives a higher epsilon value. On the other hand, the epsilon value decreases with background increase.

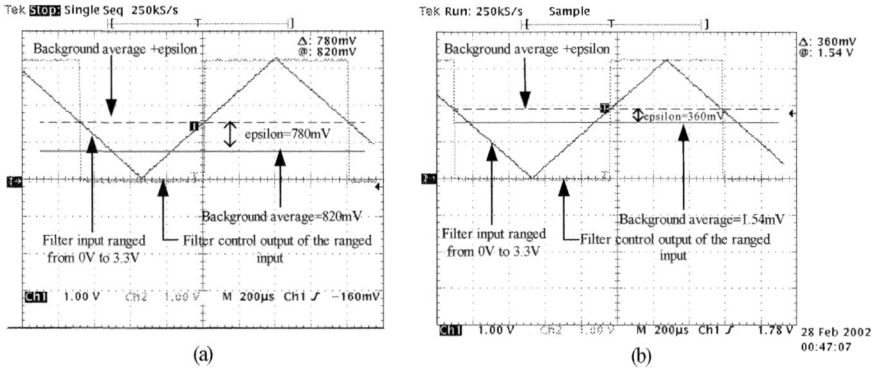

Figure 4-30. Adaptive filter response for (a) a low background and (b) a high background.

4.3.2.4.3 Measurements of the filter response and false alarm reduction

Figure 4-32 shows the filter response to four different images. The column beside each image shows the "no object" outputs for every row: white corresponds to "1" in "no object" output and black to "0". *Figure 4-32(a)* and *Figure 4-32(b)* present the same object of interest (a laser beam spot) but with two different background levels, a high background in *Figure 4-32(a)* and a low background in *Figure 4-32(b)*.

Because of the adaptive properties of the filter, the epsilon value is higher for the low backgrounds in *Figure 4-32(b)* and *(d)*, so the filtering is more aggressive and only very high values pass the filter. For the high backgrounds in *Figure 4-32(a)* and *(c)*, the epsilon value is lower and relatively lower voltage values can pass the filter. The filtering process is thus more sensitive when a high background is present and the object input

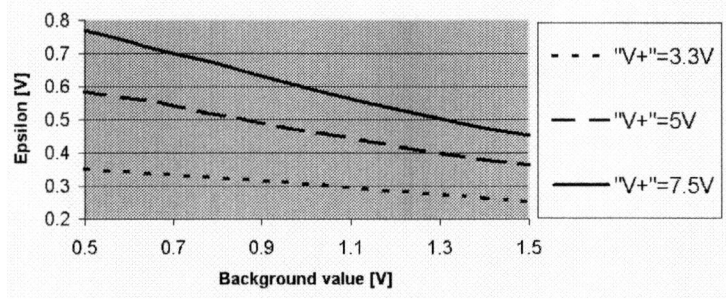

Figure 4-31. Epsilon value as function of background levels for different *V*+ values.

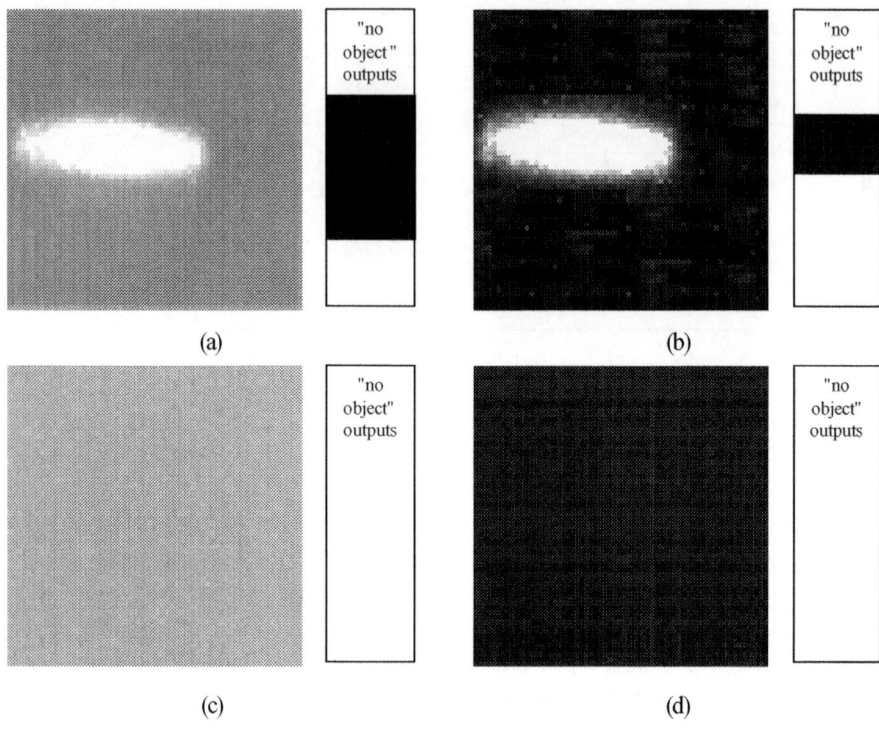

Figure 4-32. Adaptive filter response to four different images: (a) laser beam with a high background; (b) laser beam with a low background; (c) high background only; and (d) low background only.

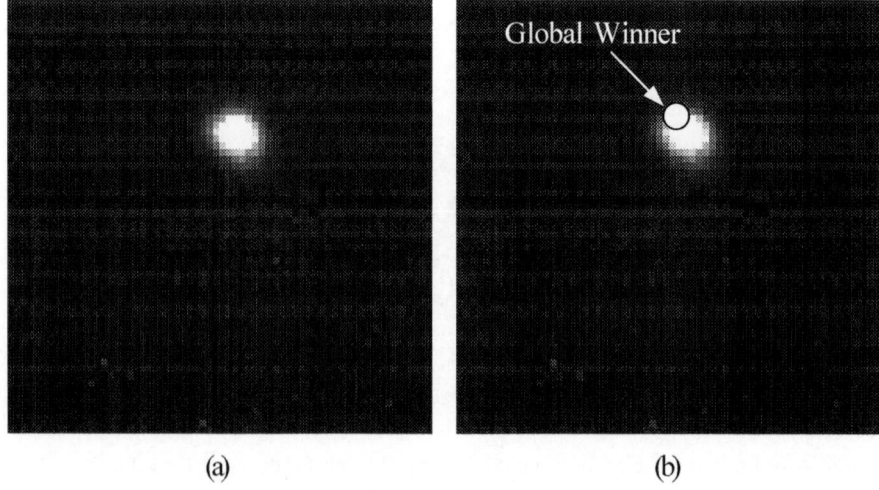

Figure 4-33. Global winner selection: (a) the processed image; and (b) the winner location.

Table 4-2. Chip attributes for the CMOS smart tracking sensor.

Technology	HP 0.5 μm
Voltage supply	3.3 V
Array size	64 × 64
Pitch width	14.4 μm
Fill factor	49%
WTA mode	Voltage
WTA resolution	40 mV
Chip size (mm)	3.5 × 4.3
Frame scanning frequency	30 Hz
Minimum power dissipation (low background without object)	~28 mW
Typical power dissipation (the laser beam is ~10% of the frame)	~36 mW
FPN (APS)	0.15%
Conversion gain (APS)	7.03 μV/e⁻
Dark response (APS output)	5.77 mV/s
Dynamic range	65 dB
Output voltage range	1.47 V

voltage is close to the background level. However, there is more freedom in filtering when a low background is present. In both cases, the object of the interest passes the filter. With a high background (small epsilon value), however, some pixels of the background succeed in passing the filter, while for a low background (high epsilon value) only the object pixels pass the filter. This effect can be seen in *Figure 4-32(a)* and *(b)*, where the "no object outputs" flags show the number of rows that succeed to pass the filter. As expected, *Figure 4-32(a)* has more "object" rows than *Figure 4-32(b)*. *Figure 4-32(c)* and *(d)* examine the false alarm reduction for low and high backgrounds. In both cases, the object of interest is not in the processed image and thus no signal passes the filter. The "no object" output is fixed on "1" when the object of interest disappears from the processed image, and therefore a false alarm is prevented for both levels of background.

4.3.2.4.4 The global winner measurements.

To examine the ability of the system to find the coordinates of the global winner, a focused laser beam was used as object of interest. *Figure 4-33(a)* shows the processed image; *Figure 4-33(b)* presents the winner as found by the system. As expected, the winner is the pixel farthest to the upper left in the first row of the object of interest.

Table 4-2 summarizes the chip specifications for the CMOS smart tracking sensor.

4.4   Summary

In this chapter, the area of CMOS imagers was briefly introduced. CMOS technologies, CCD technologies and different CMOS pixels were described and compared. The system-on-a-chip approach was presented, showing two design examples. The main advantages of CMOS imagers— low cost, low power requirements, fabrication in a standard CMOS process, low voltage and monolithic integration—rival those of traditional charge coupled devices. With the continuous progress of CMOS technologies, especially the decreasing minimum feature size, CMOS imagers are expected to penetrate into various fields such as machine vision, portable devices, security, biomedical and biometric areas, and other applications where custom sensors and smart pixels are required. More systems on chips will definitely be seen in the near future.

BIBLIOGRAPHY

[1]   E. R. Fossum, "Active pixel sensors (APS) — Are CCDs dinosaurs?" *Proc. SPIE*, vol. 1900, pp. 2–14, 1992.

[2]   D. Litwiller, "CCD vs. CMOS: facts and fiction," Photonics Spectra, pp. 154–158, January 2001.

[3]   P. Seitz, "Solid-state image sensing," in *Handbook of Computer Vision and Applications*, vol. 1, 165–222, Academic Press, 2000.

[4]   E. Fossum, "CMOS image sensors: electronic camera-on-a-chip," *IEEE Trans. Electron Devices*, vol. 44, p. 1689, 1997.

[5]   O. Yadid-Pecht and A. Belenky, "In-pixel autoexposure CMOS APS," *IEEE J. Solid-State Circuits*, vol. 38(8), pp. 1–4, August 2003.

[6]   A. Fish, D. Turchin, and O. Yadid-Pecht, "An APS with 2-dimensional winner-take-all selection employing adaptive spatial filtering and false alarm reduction," *IEEE Trans. Electron Devices*, vol. 50(1), pp. 159–165, January 2003.

[7]   G. Weckler, "Operation of p-n junction photodetectors in a photon flux integrating mode," *IEEE J. Solid-State Circuits*, vol. SC-2, p. 65, 1967.

[8]   P. Noble, "Self-scanned image detector arrays," *IEEE Trans. Electron Devices*, vol. ED-15, p. 202, 1968.

[9]   F. Andoh, K. Taketoshi, J. Yamazaki, M, Sagawara, Y. Fujita, K. Mitani, Y. Matuzawa, K. Miyata, and S. Araki, "A 250,000 pixel image sensor with FET amplification at each pixel for high speed television cameras," *IEEE ISSCC 1990 Dig. Tech. Papers*, 1990, pp. 212–213.

[10] S. K. Mendis, B. Pain, S. Kemeny, R. C. Gee, Q. Kim, and E. Possum, "CMOS active pixel image sensors for highly integrated imaging systems," *IEEE J. Solid-State Circuits*, vol. SC-32, pp. 187–198, 1997.

[11] N. Ricquier and B. Dierickx, "Pixel structure with logarithmic response for intelligent and flexible imager architectures," *Microelectron. Eng.*, vol. 19, pp. 631–634, 1992.

[12] O. Yadid-Pecht, R. Ginosar, and Y. Diamand, "A random access photodiode array for intelligent image capture," *IEEE J. Solid-State Circuits*, vol. SC-26, pp. 1116–1122, 1991.

[13] R. Hornsey, *Design and Fabrication of Integrated Image Sensors* (course), Dept. of Electrical and Computer Engineering, University of Waterloo, Waterloo, Ontario, Canada, 1998.

[14] B. Pain, *CMOS Digital Image Sensors* (SPIE course), San Jose, CA, USA, 2001.

[15] N. Marston, "Solid-state imaging: a critique of the CMOS sensor," Ph.D. Thesis, The University of Edinburgh, UK, Nov. 1998.

[16] C. Mead, "A sensitive electronic photoreceptor," in *1985 Chapel Hill Conf. on VLSI*, H. Fuchs, Ed. Rockville, MD: Computer Science Press, 1985, pp. 463–471.

[17] K. A. Boahen and A. G. Andreou, "A contrast-sensitive retina with reciprocal synapses," *Adv. Neural Information Processing*, vol. 4, pp. 762–772, 1992.

[18] C. Mead, *Analog VLSI and Neural Networks*, Addison Wesley, 1989.

[19] K. A. Boahen, "The retinotopic approach: pixel parallel adaptive amplification, filtering, and quantization," *Analog Integrated Circuits and Signal Processing*, vol. 13, pp. 53–68, 1997.

[20] T. Delbruck and C. Mead, "Adaptive photoreceptor with wide dynamic range," in *Proc. IEEE Int. Symp. Circuits and Systems*, London, UK, May 30–June 2, 1994, pp. 339–342.

[21] V. Ward, M. Syrzycki, and G. Chapman, "CMOS photodetector with built-in light adaptation mechanism," *Microelectronics J.*, vol. 24, no. 5, pp. 547–553, Aug. 1993.

[22] W. Yang, "A wide-dynamic-range, low power photosensor array," *IEEE ISSCC 1994 Dig. Tech. Papers,* 1994, pp. 230-231.

[23] R. Miyagawa and T. Kanade, "Integration time based computational image sensor," in *1995 IEEE Workshop on CCDs and Advanced Image Sensors*, Dana Point, California, USA, April 20–22, 1995.

[24] O. Yadid-Pecht, "Widening the dynamic range of pictures," in *Proc. SPIE/IS&T Symp. on Electronic Imaging: Science and Technology*, San Jose, California, Feb. 9–14, 1992, SPIE vol. 1656, pp. 374–382.

[25] O. Yadid-Pecht, "Method and apparatus for increasing the dynamic range of optical sensors," Israeli Patent number 100620, Feb. 1995.

[26] O. Yadid-Pecht, "Wide dynamic range sensors," *Opt. Eng.*, vol. 38, no. 10, pp.1650–1660, Oct. 1999.

[27] O. Yadid-Pecht and E. Fossum, "Image sensor with ultra-high linear-dynamic range utilizing dual output CMOS active pixel sensors," *IEEE Trans. Elec. Dev.*, Special Issue on Solid State Image Sensors, vol. 44, no. 10, pp. 1721–1724, Oct. 1997.

[28] T. Nakamura and K. Saitoh, "Recent progress of CMD imaging," in *Proc. 1997 IEEE Workshop on Charge-Coupled Devices and Advanced Image Sensors*, Bruges, Belgium, June 5–7, 1997.

[29] Y. Wang, S. Barna, S. Campbell, and E. R. Fossum, "A high dynamic range CMOS APS image sensor," presented at the IEEE Workshop CCD and Advanced Image Sensors, Lake Tahoe, Nevada, USA, June 7–9, 2001.

[30] D. X. D. Yang, A. El Gamal, B. Fowler, and H. Tian, "A 640 × 512 CMOS image sensor with ultra wide dynamic range floating point pixel level ADC," *IEEE ISSCC 1999 Dig. Tech. Papers*, 1999, WA 17.5.

[31] T. Lule, M. Wagner, M. Verhoven, H. Keller, and M. Bohm, "10000-pixel, 120 dB imager in TFA technology," *IEEE J. Solid State Circuits*, vol. 35, no. 5, pp. 732–739, May 2000.

[32] E. Culurciello, R. Etienne-Cummings, and K. Boahen, "Arbitrated address event representation digital image sensor," *IEEE ISSCC 2001 Dig. Tech. Papers*, (Cat. No. 01CH37177), Piscataway, NJ, USA, 2001, pp. 92–3.

[33] L. G. McIlrath, "A low-power, low-noise, ultrawide-dynamic-range CMOS imager with pixel-parallel A/D conversion," *IEEE J. Solid State Circuits*, vol. 36, no. 5, pp. 846–853, May 2001.

[34] T. Hamamoto and K. Aizawa, "A computational image sensor with adaptive-pixel-based integration time," *IEEE J. Solid State Circuits*, vol. 36, no. 4, pp. 580–585, Apr. 2001.

[35] O. Yadid-Pecht and A. Belenky, "Autoscaling CMOS APS with customized increase of dynamic range," in *Proc. IEEE ISSCC*, San Francisco, CA, USA, February 4–7, 2001, pp. 100–101.

[36] O. Yadid-Pecht, "The automatic wide dynamic range sensor," in *1993 SID Int. Symp.*, Seattle, WA, USA, May 18–20, 1993, pp. 495–498.

[37] O. Yadid-Pecht, B. Pain, C. Staller, C. Clark, and E. Fossum, "CMOS active pixel sensor star tracker with regional electronic shutter," *IEEE J. Solid State Circuits*, vol. 32, no. 2, pp. 285–288, Feb. 1997.

[38] O. Yadid-Pecht, E. R. Fossum, and C. Mead, "APS image sensors with a winner-take-all (WTA) mode of operation," *JPL/Caltech New Technology Report*, NPO 20212.

[39] Z. S. Gunay and E. Sanches-Sinencio, "CMOS winner-take-all circuits: a detail comparison," in *IEEE ISCAS'97*, Hong Kong, June 9–12, 1997, pp. 41–44.

[40] J. Lazzaro, S. Ryckebusch, M. A. Mahowald, and C.A. Mead, "Winner-tale-all networks of O(n) complexity," in *Advances in Neural Information Processing Systems*, Vol. 1, D. S. Touretzky, Ed. San Mateo, CA: Morgan Kaufmann, 1989, pp. 703–711.

[41] J. A. Startzyk and X. Fang, "CMOS current mode winner-take-all circuit with both excitatory and inhibitory feedback," *Electronics Lett.*, vol. 29, no. 10, pp. 908–910, May 13, 1993.

[42] G. Indivery, "A current-mode hysteretic winner-take-all network, with excitatory and inhibitory coupling," *Analog Integrated Circuits and Signal Processing*, vol. 28, pp. 279–291, 2001.

[43] T. Serrano and B. Linares-Barranco, "A modular current-mode high-precision winner-take-all circuit." *IEEE Trans. Circuits and Systems II*, vol. 42, no. 2, pp. 132–134, Feb. 1995.

[44] G. Indiveri, "Neuromorphic analog VLSI sensor for visual tracking: circuits and application examples," *IEEE Trans. Circuits and Systems II*, vol. 46, no. 11, pp. 1337–1347, Nov. 1999.

[45] S. P. DeWeerth and T. G. Morris, "CMOS current mode winner-take-all circuit with distributed hysteresis," *Electronics Lett.*, vol. 31, no. 13, pp.1051-1053, 22 June 1995.

[46] D. M. Wilson and S. P. DeWeerth, "Winning isn't everything," in *IEEE ISCAS'95*, Seattle, WA. USA, 1995, pp. 105–108.

[47] R. Kalim and D. M. Wilson, "Semi-parallel rank-order filtering in analog VLSI," in *IEEE ISCAS'99*, Piscataway, NJ, USA, 1999, vol. 2, pp. 232–5.

[48] N. Donckers, C. Dualibe, and M. Verleysen, "Design of complementary low-power CMOS architectures for loser-take-all and winner-take-all," *Proc. 7th Int. Conf. on Microelectronics for Neural, Fuzzy and Bio-Inspired Systems*, Los Alamitos, CA, USA: IEEE Comput. Soc., pp. P360–5, 1999.

[49] T. G. Moris and S. P. DeWeerth, "A smart-scanning analog VLSI visual-attention system," *Analog Integrated Circuits and Signal Processing*, vol. 21, pp. 67–78, 1999.

[50] Z. Kalayjian, J. Waskiewicz, D. Yochelson, and A. G. Andreou, "Asynchronous sampling of 2D arrays using winner-takes-all arbitration," *IEEE ISCAS'96*, New York, USA, 1996, vol. 3, pp. 393–6.

[51] T. G. Moris, C. S. Wilson, and S. P. DeWeerth, "Analog VLSI circuits for sensory attentive processing," *IEEE Int. Conf. on Multisensor Fusion and Integration for Intelligent Systems*, 1996, pp. 395–402.

[52] T. G. Morris, T. K. Horiuchi, and P. DeWeerth, "Object-based selection within an analog visual attention system," *IEEE Trans. Circuits and Systems II, Analog and Digital Signal Processing*, vol. 45, no. 12, pp. 1564–1572, Dec. 1998.

[53] A. Fish and O. Yadid-Pecht, "CMOS current/voltage mode winner-take-all circuit with spatial filtering," *Proc. IEEE ISCAS'01*, Sydney, Australia, May 2001, vol. 2, pp. 636–639.

Chapter 5

FOCAL-PLANE ANALOG IMAGE PROCESSING

Matthew A. Clapp, Viktor Gruev, and Ralph Etienne-Cummings
Department of Electrical and Computer Engineering
Johns Hopkins University
Baltimore, MD 21218, USA

Abstract: In this chapter, three systems for imaging and visual information processing at the focal plane are described: current-mode, voltage-mode and mixed-mode image processing. It is demonstrated how spatiotemporal image processing can be implemented in the current and voltage modes. A computation-on-readout (COR) scheme is highlighted; this scheme maximizes pixel density but still allows multiple processed images to be produced in parallel. COR requires little additional area and access time compared to a simple imager, and the ratio of imager to processor area increases drastically with scaling to smaller-feature-size CMOS technologies. In some cases, it is necessary to perform computations in a pixel-parallel manner while still retaining the imaging density and low-noise properties of an APS imager. Hence, an imager that uses both current-mode and voltage-mode imaging and processing is presented. The mixed-mode approach has some limitations, however, and these are described in detail. Three case studies are used to show the relative merits of the different approaches for focal-plane analog image processing.

Key words: Focal-plane analog image processing, current-mode image processing, voltage-mode image processing, spatiotemporal image filtering, time difference imaging, motion detection, centroid localization, CMOS imagers, analog VLSI.

5.1 Introduction

Integrating CMOS active pixel sensors (APS) [1–2] with carefully chosen signal processing units has become a trend in the design of camera-on-a-chip systems [3–4]. The benefits of low cost and low power from the emerging CMOS imaging technology have encouraged various research directions in creating image processing sensors. In the early 1990s, Mead initiated the concept of including processing capabilities at the focal plane of imagers with his neuromorphic design paradigm [5]. Mead's neuromorphic approach has inspired many other engineers to mimic the receptive fields (i.e., the spatiotemporal impulse responses) of human retinal cells. These cells perform convolution of the center surround kernel with the incident image [1–9]. Such retinomorphic systems are ideal for fast early processing. The inclusion of processing circuitry within the pixels, however, prevents the retinomorphic systems from acquiring high-resolution images. These systems are also optimized to realize specialized spatiotemporal processing kernels by hardwiring the kernels within the pixels, thus limiting their use in many algorithms. Boahen and Andreou introduced increased flexibility by using external biases to control the size of the hardwired convolution kernels [8]. Shi recently presented a specialized imager for computing various Gabor filtered images [10], in which the convolved images were used to detect object orientation. Although focal plane processing with programmable cellular networks [1–14] and switch capacitor networks [15] provide another direction for solving the programmability problem, they have resolution limitations similar to those of the retinomorphic approach.

Digital focal-plane implementation is another approach to fast and high-precision image processing. This approach requires analog-to-digital conversion (ADC) and one or many digital processing units (typically a CPU/DSP core). In these architectures, multi-chip systems or complicated digital system-on-chip units are needed. Typically the imaging and ADC is performed on one chip, whereas the computation is performed in the digital domain on a second chip [16–17]. High power consumption, complex inter-chip interconnections and poor scalability are the usual limitations. A single chip solution has been discussed where the imaging, ADC and digital processing are included at the focal plane [18]. However, only a very small percentage of that chip is used for imaging.

Integrating simultaneous spatial and temporal processing on the same substrate allows more versatility for computer vision applications. The computational time for computer vision algorithms can be drastically improved when spatial and temporal processing are done at the focal plane. This information is then presented to the CPU together with the intensity image. Hence, the first layer of computation can be performed at the focal plane, freeing computational resources for other higher-level tasks.

In this chapter, three systems that perform imaging and visual information processing at the focal plane are described: current-mode, voltage-mode and mixed-mode image processing. It is demonstrated how spatiotemporal image processing can be implemented in the current and voltage modes. A computation-on-readout (COR) scheme is highlighted; this scheme maximizes pixel density but still allows multiple processed images to be produced. It requires little additional area and access time compared to a simple imager. In some cases, it is necessary to perform computation in a pixel-parallel manner while still retaining the imaging density and low-noise property of an APS imager. Hence, an imager that uses both current-mode and voltage-mode imaging and processing is presented. The mixed-mode approach has some limitations, however, and these are described in detail. Three case studies are used to show the relative merits of the different approaches for focal-plane analog image processing.

5.2 Current-domain image processing: the general image processor

5.2.1 System overview

The three main components of the general image processor (GIP) are (1) a photopixel array of 16 rows by 16 columns, (2) three vertical and three horizontal scanning registers, and (3) a single processing unit. The three vertical and three horizontal scanning registers select several groups of single or multiple pixels within a given neighborhood of the photo array. The non-linearly amplified photocurrent values of the selected pixels are then passed to the processing unit, where convolutions with the desired filter are computed. The processing unit, which consists of four identical but independent subprocessors, is implemented with digitally controlled analog multipliers and adders. The multipliers and adders scale each of the pixel photocurrents according to the convolution kernel being implemented (see *Figure 5-1*). The final output of the processing unit is a sum of scaled photocurrents from the selected neighborhood. The independent groups of pixels can be combined in various ways, allowing for the realization of various complicated separable and non-separable filters. Each of the four subprocessors can be independently programmed in parallel, allowing for different spatiotemporal convolutions to be performed on the incident image in parallel.

5.2.2 Hardware implementation

The pixel is composed of a photodiode, a non-linear photocurrent amplification circuit, a sample-and-hold image delay element, and pixel selection switches (see *Figure 5-2*). The photodiode is implemented as the source diffusion extension of an NFET load transistor, M1. The photocurrent

© V. Gruev and R. Etienne-Cummings, "Implementation of steerable spatiotemporal image filters on the focal plane," IEEE Trans. Circuits and Systems-II, vol. 49, no. 4, pp. 233–244, Apr. 2002 (partial reprint).

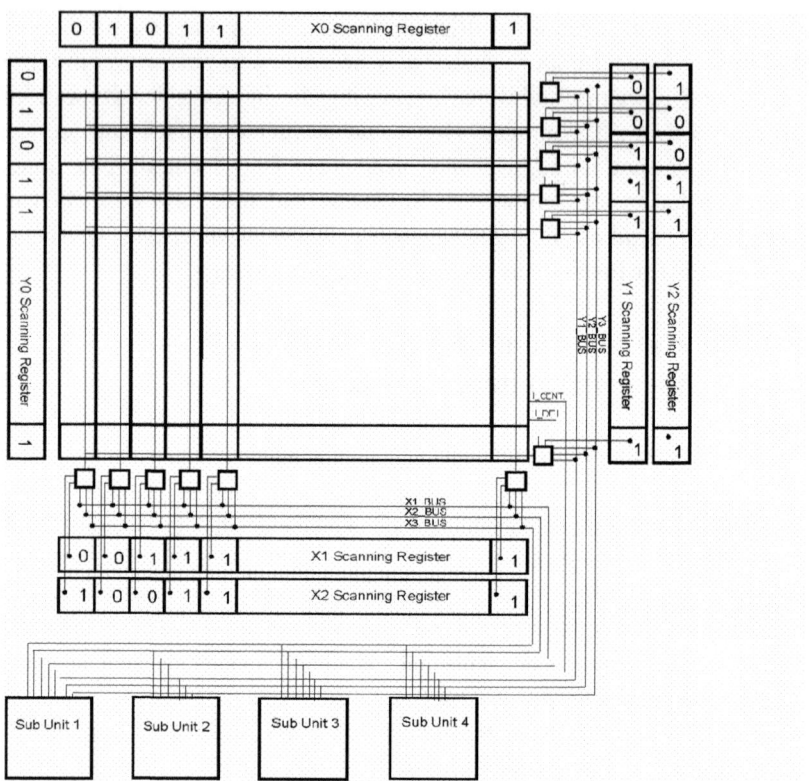

Figure 5-1. Overview of the general image processor (GIP) architecture.

and the non-linear resistance of M1 produce voltage at the source of M1. Transistors M2, M3, M4, and M11 (after sample-and-hold) behave as transconductors and transform the source voltage of M1 back to current with a non-linear gain. This circuit magnifies the photocurrent by up to three orders of magnitude.

The load transistor M1 can be operated in an integrative or non-integrative/fixed-bias mode. In the integrative mode, a pulse train is applied to the gate of the M1 transistor, alternating between the integrative and reset interval of the transistor. During the reset interval, the gate of the M1 transistor is pulled to V_{dd} or higher, charging the photodiode and source node of this transistor to $\geq V_{dd} - V_{Tn(sb)}$.

With the four output transistors and the photodiode, a capacitance of ~ 65 fF is expected at the source of M1. When coupled with the g_m of M1 ($\sim 2 \times 10^{-7}$ mhos with a 1 nA photocurrent), this provides a reset time constant of ~ 0.3 µs. Clearly the reset time would be problematic if

© V. Gruev and R. Etienne-Cummings, "Implementation of steerable spatiotemporal image filters on the focal plane," *IEEE Trans. Circuits and Systems-II*, vol. 49, no. 4, pp. 233–244, Apr. 2002 (partial reprint).

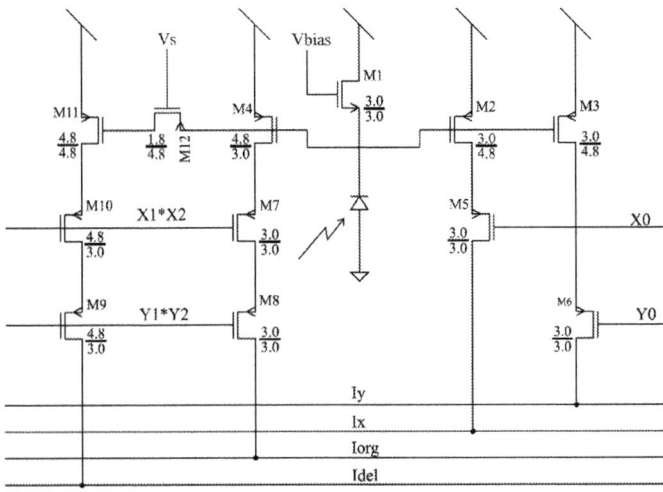

Figure 5-2. Organization of the GIP photopixel. The photocurrent is amplified and mirrored four times. A sample-and-hold circuit is included in the pixel to provide the delayed image.

difference or correlated double sampling (DDS or CDS) were used for $V_{Tn(sb)}$ variation cancellation while scanning the array above 3 MHz. No DDS was attempted here.

During the integration period, the gate voltage of M1 is pulled down to V_{ss}, which turns it off. The photocurrent from the diode discharges the floating source node of M1. In the non-integrative mode, a constant bias voltage is applied to the gate of M1, so that M1 becomes a non-linear resistive load. The currents produced by M2, M3, M4, and M11 are scanned and passed to the processing unit for further processing.

The transistors M2, M3, M4, and M11 provide non-linearly amplified photocurrents. Three of the amplified photocurrents (I_x, I_y, and I_{org}) are used for spatial processing. The fourth is used for temporal processing, and is passed through a sample-and-hold circuit that stores the amplified photocurrent for many frame times. In the integrative mode, the sampling switch is opened at the end of the integration cycle, thus holding the charge on the gate of M11. In the fixed-bias mode, the sampling switch is pulsed when a new stored image is required. In the former case, two distinct integration cycles are required for each processing pass, whereas in the latter case, processing can be done in each scanning pass. The sampling switch established by M12 can also act as an electronic shutter that controls the exposure time of the pixel; this is in addition to controlling the timing when the source voltage of M1 is sampled to the gate capacitance of M11. The gate capacitance of M11 acts as a holding node, and has been measured to maintain the output current for seconds in low light. The leakage currents of the drain/source diodes of M12 limit the time for storing the photocurrent, but this leakage can be reduced using shielding. (The process data indicates a diode junction leakage of ~ 2 fA/μm^2 under unknown reverse-bias voltage. The capacitance of M11 is ~ 27 fF and the leakage diode area is 5.8 μm^2; hence, the expected leakage rate is ~ 2.3 s/V. A longer time is observed in reality.) Some temporal image smear is expected due to charge redistribution at M11 during sampling. This is acceptable and perhaps beneficial for the delayed image, however, since it low-pass filters the stored image.

The six scanning registers are used to select groups of pixels and direct their photocurrents to the eight global current buses. The selection of the groups of pixels is accomplished into two phases. These two phases are applied to both vertically and horizontally (in parentheses below) directed currents. In the first phase, the top (left) scanning register selects all the pixels in the given columns (rows) of interest (see *Figure 5-1*). The photocurrent values of these pixels are then summed horizontally (vertically), providing the summed photocurrent values on each of the 16 rows (columns). In the second phase, the right (bottom) scanning registers select three of the previously activated 16 rows (columns) and direct each one of them to a separate vertical (horizontal) bus. This phase is achieved through a single analog multiplexer per row (column), where the control bits of the multiplexer are specified by the two registers on the right (bottom). Since there is a total of three global vertical and three global horizontal buses on the right and bottom of the photo array, respectively, a total of six different groups of pixels are selected and passed to the processing unit for further processing.

The bottom two registers and the two right registers are used to select one additional group of pixels. The selection of this group of the pixels is achieved in two steps. In the first step, bit slices of the bottom (right) two registers are NANDed. Subsequently, the NAND results activate PFET switches for the I_{del} (delayed image) and I_{org} (original image) currents for the pixels of interest (see *Figures 5-1* and *5-2*). These two currents are passed to the processing unit for spatiotemporal processing.

The virtual ground (VG) circuit is designed using the four-transistor current conveyor circuit shown in *Figure 5-3*. Virtual ground circuits are used to mask the large capacitance of the current buses. The impedance at the input of the VG circuit can be easily derived to be $\sim 2/g_{m(PFETs)}$, assuming the NFETs and PFETs have the same dimensions. Assuming a single pixel is selected per current output line, a quiescent pixel (dark) current of ~ 1 μA

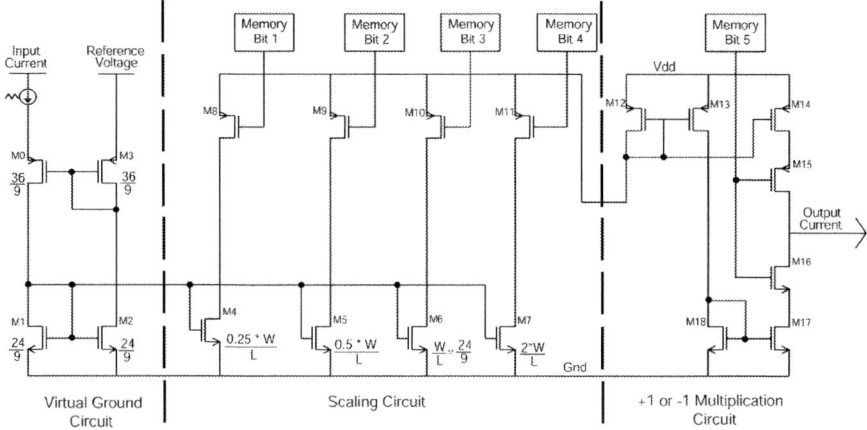

Figure 5-3. Schematics of the virtual ground circuit and scaling circuits.

flows into the VG circuit (see *Figure 5-9*) and the *W/L* ratio of all the transistors is 10, giving an input impedance of ~ 200 kΩ for this CMOS process. In the light or with larger kernels, more pixels are selected so that the quiescent current is larger and leads to a smaller VG input impedance. The line capacitance, primarily determined by the drain diffusion capacitance of the readout switches, is ~ 2 fF/pixel in a row or column. With 16 pixels on each row, the VG response time is 6.4 ns; with 1000 pixels per row, it is 400 ns. Hence, it is easy to obtain the required VG response time constant by increasing the quiescent current.

Nine virtual ground circuits are used in this architecture. Eight of these are used for the eight global current buses that bring the pixel currents to the processing unit. The last virtual ground circuit connects together all the current buses (both horizontal and vertical) that are not selected by the scanning registers and then connects them to a reference voltage. This virtual ground circuit keeps the voltage of all the bus currents at a fixed value, even when they are not selected. Hence, the pixel current buses are precharged at all times. During readout, the pixel currents are simply connected to the virtual ground circuit or to the reference voltage.

The processing unit is a digitally controlled analog processing unit consisting of four subunits. The subunits are identical in structure, each with a digital control memory of 40 bits as well as analog scale and add circuits. Each of the eight input currents are mirrored four times and then passed to the subprocessors for individual computation. The digital memory assigns a 5-bit signed-magnitude control word per current, which specifies the kernel coefficient for each current (see *Figure 5-3*). The coefficient can vary within a range of ±3.75 (in increments of 0.25). The appropriate weight factors vary depending on the given mask of interest. After each current is weighted by the appropriate factor, all currents are summed together to produce the desired processed image.

5.2.3   Algorithm

To capitalize on the power of the parallel processing capabilities of the GIP, the sizes of the pixel groups are kept small. Minimizing the number of pixels per group maximizes the number of independent kernels that can be implemented in parallel. Ideally, if every pixel value for a given neighborhood is available to the processing unit, the kernels can be completely general (i.e., every pixel can be given its own coefficient). This is not possible in this architecture without using a large number of pixel current copies. Since pixel size and spacing would be excessive due to a large number of current routing lines, a completely general implementation would be impractical. The trade-off between generality and pixel size must be taken into an account in designing the GIP. Hence, our design allows for computation of variable sizes of kernels based on a 3×3 canonical model, where eight unique coefficients can be applied to the nine pixels. The distribution of these coefficients depends on the configuration of the switches (or registers) for selection and routing.

To illustrate this point, a 3×3 block can be considered in which the scanning registers are loaded with the bit patterns shown in *Figure 5-1*. After this is completed, the global current buses will carry the groups of pixel currents that are described in equations (5.1a) through (5.1h):

$$I_{X1} = I_{(1,1)} + I_{(1,3)} \tag{5.1a}$$

$$I_{X2} = I_{(2,1)} + I_{(2,3)} \tag{5.1b}$$

$$I_{X3} = I_{(3,1)} + I_{(3,3)} \tag{5.1c}$$

$$I_{Y1} = I_{(1,1)} + I_{(3,1)} \tag{5.1d}$$

$$I_{Y2} = I_{(1,2)} + I_{(3,2)} \tag{5.1e}$$

© V. Gruev and R. Etienne-Cummings, "Implementation of steerable spatiotemporal image filters on the focal plane," *IEEE Trans. Circuits and Systems-II*, vol. 49, no. 4, pp. 233–244, Apr. 2002 (partial reprint).

Table 5-1. Summary of convolution kernel construction.

x edges	$-I_{X1} - I_{Y2} - I_{Y3} + 2I_{X2} + 2I_{org}$
y edges	$-I_{X1} - I_{X2} - I_{X3} + 2I_{Y2} + 2I_{org}$
45° edges	$I_{X1} - I_{Y3}$
135° edges	$I_{X1} - I_{Y1}$
Gaussian	$2I_{org} + I_{X2} + I_{Y2}$
Laplacian	$-I_{Y2} - I_{X2} + 4I_{org}$
Rectangular smooth	$(I_{X1} + I_{X2} + I_{X3} + I_{Y2} + I_{org})/9$
Temporal derivative	$I_{org} - kI_{del}$

$$I_{Y3} = I_{(1,3)} + I_{(3,3)} \tag{5.1f}$$

$$I_{org} = I_{(2,2)} \tag{5.1g}$$

$$I_{del} = I_{(2,2,t-1)} \tag{5.1h}$$

Other combinations are possible, but the grouping presented will yield the maximum number of independent kernels processed in parallel. For large kernels, the current $I_{(i,j)}$ can be the sum of multiple individual pixel currents, such as those of a 3×3 subregion. Using these currents as the basis, various kernels can be constructed. *Table 5-1* gives some examples of convolution kernels realized with the pixel grouping presented in equations (5.1).

A general convolution with an $M \times N$ kernel is given in equation (5.2). Four such convolutions (with different coefficients) can be executed in parallel.

$$I_{out} = \sum_{i=1}^{3}(a_i I_{xi} + b_i I_{yi}) + cI_{org} + dI_{del} \tag{5.2a}$$

$$\{a_i, b_i, c, d\} = n/4 \quad for \quad -15 \le n \in I \le +15 \tag{5.2b}$$

$$I_{\{x,y\},i} = \sum_{k,l}^{M,N} e_{kl} I_{ph}(k,l) \quad where \quad e_{kl} \in \{0,1\} \tag{5.2c}$$

© V. Gruev and R. Etienne-Cummings, "Implementation of steerable spatiotemporal image filters on the focal plane," IEEE Trans. Circuits and Systems-II, vol. 49, no. 4, pp. 233–244, Apr. 2002 (partial reprint).

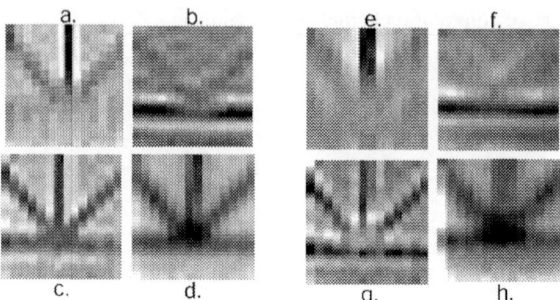

Figure 5-4. Spatial convolution results from the GIP. *(a)* Horizontal edge detection obtained from 3 × 3 mask. *(b)* Vertical edge detection obtained from 3 × 3 mask. *(c)* 2D edges obtained from a 3 × 3 Laplacian mask. *(d)* Smoothed image obtained from a 3 × 3 Gaussian mask. *(e)* Horizontal edge detection obtained from 5 × 5 mask. *(f)* Vertical edge detection obtained from 5 × 5 mask. *(g)* 2D edges obtained from a 5 × 5 Laplacian mask. *(h)* Smoothed image obtained from a 5 × 5 Gaussian mask.

5.2.4   Results

The GIP was tested using a programmed bit pattern discussed later in this chapter. *Figure 5-4* shows examples of the outputs of the chip when the incident image is convolved with the convolution kernels in *Figure 5-5*. Images *(a)* to *(d)* are computed using 3 × 3 convolution kernels, whereas images *(e)* to *(h)* are computed using 5 × 5 convolution kernels. A 5 × 5 kernel is identical to a 3 × 3 kernel except that a 2 × 2 subregion replaces the pixels that are off the central axis of the kernel (see *Figure 5-5*). Similar techniques are used to construct larger kernels. The top row of images in *Figure 5-4* shows the vertical and horizontal edge detection images

Figure 5-5. Convolution kernels coefficients used for spatial processing in *Figure 5-4*.

© V. Gruev and R. Etienne-Cummings, "Implementation of steerable spatiotemporal image filters on the focal plane," *IEEE Trans. Circuits and Systems-II*, vol. 49, no. 4, pp. 233–244, Apr. 2002 (partial reprint).

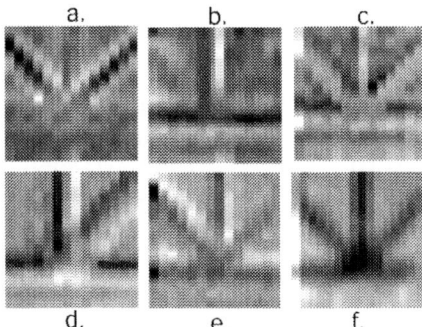

Figure 5-6. Some complex and non-separable convolution kernels. *(a)* Diagonal edge detection. *(b)* Horizontal and vertical edge detection. *(c)* High-pass filtered image. *(d)* –45° edge detection. *(e)* +45° edge detection. *(f)* Low-pass filtered image.

respectively, computed by the two kernel sizes. The bottom row of images shows the Laplacian and Gaussian images respectively. As expected, the vertical black line in images *(a)* and *(f)* is not visible in the horizontal edge images, *(b)* and *(g)*. Both horizontal and vertical edges are visible in the Laplacian image, whereas the Gaussian image provides a smooth version of the image.

Figure 5-6 shows further examples of images filtered by various complex, non-separable, or rotated filters. The kernels are shown in *Figure 5-7*. Image *(a)* is the result of convolution with a spatial mask that only computes the diagonal edges of the image, whereas the kernel for image *(b)* highlights a combination of horizontal and vertical edges while suppressing the diagonal

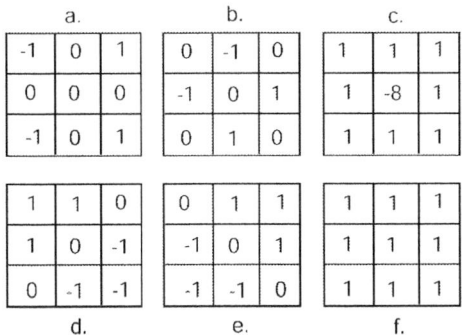

Figure 5-7. Non-separable convolution kernels coefficients used for spatial processing in *Figure 5-6*.

© V. Gruev and R. Etienne-Cummings, "Implementation of steerable spatiotemporal image filters on the focal plane," IEEE Trans. Circuits and Systems-II, vol. 49, no. 4, pp. 233–244, Apr. 2002 (partial reprint).

152

Chapter 5

Table 5-2. Summary of GIP characteristics.

Technology	1.2 μm Nwell CMOS
No. of transistors	6000
Array size	16 × 16
Pixel size	30 μm × 30 μm
Fixed-pattern noise (STD/mean)	2.5% (average)
Fill factor	20%
Dynamic range	1–6000 lux
Frame rate range	DC–400 kHz
Kernel size range	From 2 × 2 up to whole array
Kernel coefficients	±3.75 by 0.25
Coefficient of precision (STD/mean)	Intraprocessor: <0.5%
	Interprocessor: <2.5%
Temporal delay (holding time)	1% decay in 150 ms @ 800 lux
	1% decay in 11ms @ 8000 lux
Power consumption (V_{dd} = 5 V @ 800 lux)	5 × 5 array: 1 mW @ 20 kfps
Computation rate (add and multiply)	5 × 5 array: 1 GOPS/mW @ 20 kfps

edges. Images *(d)* and *(e)* show –45° and +45° edge detection respectively, and image *(f)* shows a low-pass filtered version of the image. Furthermore, the kernels for images *(a)* and *(c)* compute second-order derivatives (oriented Laplacian operators) of the intensity, whereas images *(b)*, *(d)*, and *(e)* are first-order derivatives (oriented gradient operators). Hence, the sign of the slope of the intensity gradient is observable in the latter images, but not in the former. The orientation selectivity of the 2D edge detectors is clearly visible in these figures. Other complex and non-separable filters can also be implemented with the GIP.

Table 5-2 shows the general characteristics of the chip. A much larger array is possible with no impact on the performance of the scanning circuit and processor unit. In a larger array, most of the additional area will be in the photo array because the overhead for the scanning and processing circuits will be similar to that required in this chip. The circuitry of the processing unit is independent of the photo array size and will be the same as in the smaller photo array, although it will be redesigned to handle large convolution kernels. The current handling capability must be increased. Similar digital processors typically use more than 50% of the area budget in the processing units [19].

The amplified photoresponses of many randomly distributed pixels in the fixed-bias mode (V_{bias} = 5 V) are shown in *Figure 5-8(a)*. The light intensity is varied over six orders of magnitude. For very low light intensity, both M1 and M2 transistors in *Figure 5-2* operate in weak inversion mode, resulting in a linear relation between the light intensity and the amplified

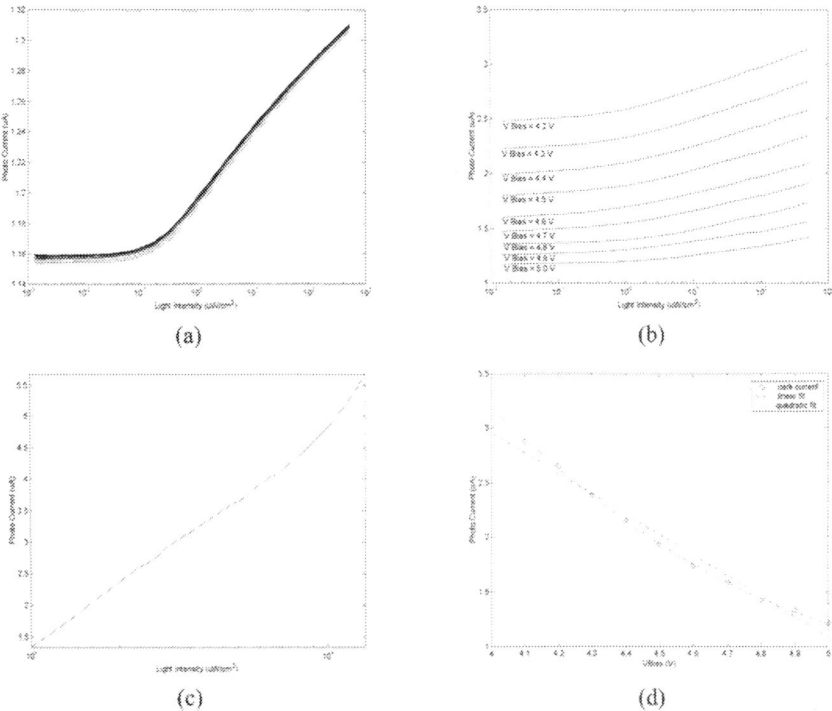

Figure 5-8. Photo response of the GIP pixel operating in fixed-bias mode. *(a)* Low-to-medium light response of 50 randomly distributed pixels with $V_{bias} = 5$ V. *(b)* Low-to-medium light response of a pixel for various V_{bias} values. *(c)* Medium-to-bright light response of the pixel. *(d)* Output dark current for various V_{bias} values.

photocurrent. This dependence is described by equation (5.3), but is not observed in *Figure 5-8*. This is because dark current in the photodiode produced a large enough gate-to-source voltage drop in M1 (~ 0.96 V) due to the body effect ($\{V_{TON}, \gamma_N\} = \{0.6$ V, 0.62 V$^{-2}\}$) and $\{V_{TOP}, \gamma_P\} = \{0.7$ V, 0.4 V$^{-2}\}$) to operate M2 above threshold. Models for transistors in below- and above-threshold states can be found in [20]. A minimum (dark) current of 1.16 μA flows in M2; this current can be reduced or eliminated by raising the gate voltage of M1 above V_{dd}.

Equation (5.4) shows the dependence of the output current on photocurrent when M1 is in weak inversion and M2 is in strong inversion and saturated. This equation indicates that the output current has a log dependence for small photocurrents and a log-squared dependence for large photocurrents. The log dependence for small photocurrents is observed in *Figure 5-8(a)*. The slope of the curve is determined by the bias gate voltage of M1 and the difference in threshold voltages of M1 and M2. It should be noted that that the difference in threshold voltages decreases with increasing light because V_s of M1 approaches its V_b (i.e., V_{ss}). This is also observable in *Figure 5-8* as a decrease in slope with increasing light intensity. In *Figure 5-8(b)*, a range of gate bias voltages for M1 is shown. As predicted by equation (5.4), the slope of the curve increases as $V_{dd} - V_{bias}$ increases. *Figure 5-8(c)* shows the transition from log to log-squared dependence as the light intensity is increased drastically. The imager will be used in the logarithmic mode to widen the dynamic range of the pixel. It is also true that biological photoreceptors exhibit logarithmic responses [5]; the logarithmic relationship has profound influences not only on the dynamic range of the pixel, but also on grayscale and color scene analysis. *Figure 5-8(d)* shows the pixel output current in the dark as V_{bias} is varied. The expected quadratic dependence observed.

As the photocurrent increases further, both M1 and M2 transistors operate in a strong inversion mode, resulting in a linear relation between the light intensity and the amplified photocurrent. This dependency is given by equation (5.5) and was not reached in our experiments.

$$i_{out} \propto \frac{(\frac{W}{L})_{M2}}{(\frac{W}{L})_{M1}} i_{photo} \tag{5.3}$$

$$i_{out} = \frac{k_p}{2}(\frac{W}{L})_{M2}\left((Vdd - Vbias + VTn(sb) - VTOp) + (nV_t)\ln[\frac{i_{photo}}{(\frac{W}{L})_{m1}I_o}]\right)^2$$

$$\approx \frac{k_p}{2}(\frac{W}{L})_{M2}\left(V_{const}^2 + V_{const}(nV_t)\ln[\frac{i_{photo}}{(\frac{W}{L})_{m1}I_o}] + \left\{(nV_t)\ln[\frac{i_{photo}}{(\frac{W}{L})_{m1}I_o}]\right\}^2\right) \tag{5.4}$$

$$i_{out} \propto \frac{k_p}{k_n}\frac{(\frac{W}{L})_{M2}}{(\frac{W}{L})_{M1}} i_{photo} \tag{5.5}$$

where V_t is the thermal voltage, I_0 and $k_{\{p,n\}}$ are process parameters, i_{photo} is the photocurrent passing through transistor M1, and i_{out} is the amplified

© V. Gruev and R. Etienne-Cummings, "Implementation of steerable spatiotemporal image filters on the focal plane," *IEEE Trans. Circuits and Systems-II*, vol. 49, no. 4, pp. 233–244, Apr. 2002 (partial reprint).

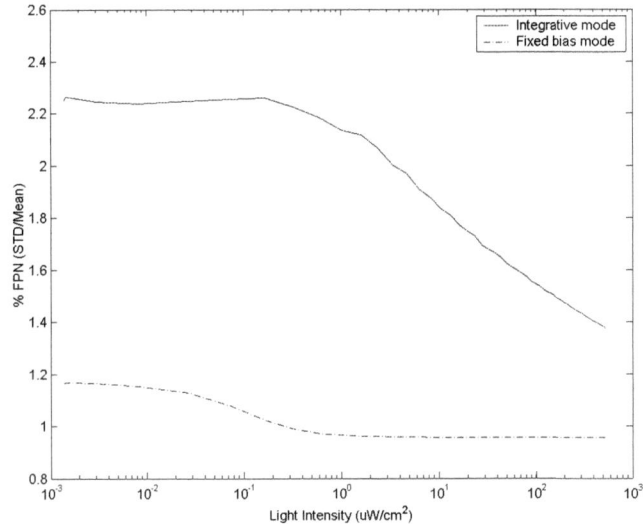

Figure 5-9. Fixed-pattern noise of the GIP pixel operating in integrative and non-integrative modes.

photocurrent passing through transistor M2.

The fixed-pattern noise (FPN) characteristics for integrative and non-integrative modes are shown in *Figures 5-8* and *5-9*. In *Figure 5-8(a),* the response of many pixels under fixed-bias operation is shown, and the variation at each light intensity is visible. As expected, the variations appear as offsets to the pixel currents. The magnitude of the variation remains relatively constant with light intensity; hence, the % FPN (STD/mean) will decrease with light intensity, as seen in *Figure 5-9*.

The noise characteristics for low light intensity are better when the pixels operate in the integrative mode. In this mode, transistor M1 is either turned off during the integration period or is operated in a strong inversion mode during the reset period. To reduce the variations in the voltage at the start of integration (i.e., the reset voltage), the pixel is reset by driving the gate of M1 above V_{dd}. With a sufficiently high gate voltage [i.e., greater than $(V_{dd} + V_{Tn(sb)})$], the reset voltage will be V_{dd} for all pixels. In this case, the fixed-pattern noise depends primarily on the M2 transistor and it is not affected by transistor M1. When M1 operates in fixed-bias mode, both M1 and M2 operate in a weak inversion mode in low light intensities. Hence, the variations in M1 and M2 (which are uncorrelated because the transistors are different types) will contribute to the noise characteristics. In low light, FPN for the integrative case is found to be approximately half that of the fixed-bias case (1.2% vs. 2.3%). It is well known that currents of the transistors match better (in terms of σ_{Ids}/μ_{Ids}) in strong inversion than in weak inversion because the mean current is large in the former case [21]; M1 and M2 operate in strong inversion in bright lights under both integrative and fixed-bias operation. For the integrative case, the output current is larger than in the fixed-bias case because the gate voltage of M2 can be large. Consequently, the STD/mean for the integrative mode quickly (first exponentially, then quadratically) reaches its minimum of less than 1%. In typical indoor lighting, the FPN will limit the image signal-to-noise ratio (SNR) to ~ 6 bits and ~ 7 bits for the fixed-bias and integrative modes, respectively. Correlated or difference double sampling can be used to improve the FPN [22].

The scaling precision of the processing unit is shown in *Figure 5-10*. In *Figure 5-10(a)*, the imager is stopped on a pixel and a processing unit is programmed with all 32 coefficients under nine ambient light levels. The plot shows that the weights vary about the ideal $x = y$ line, but the variation is quite small. *Figure 5-10(b)* shows that the variation in terms of STD/mean is less than 0.5% (~ 8 bits) across coefficients. Hence, within one processing unit, the precision of computation is approximately 8 bits. *Figure 5-10(c)* compares the matching across the four processing units for one light intensity. *Figure 5-10(d)* shows that the precision across the processing unit is poorest for the smallest coefficients. This is again consistent with the expectation that for smaller currents, STD/mean becomes larger if the STD remains relatively unchanged. The matching across processing units is approximately 2.5% (~ 5 bits). Better layout practices can be used to improve the matching of processing units.

The total power consumption for computing four convolved images with 5×5 spatiotemporal kernels is 1 mW with 5 V supplies and 10 μW/cm^2 ambient light. The power consumption can be easily decreased by modifying the gain of the pixels and by decreasing the power supply voltage. Convolving the incident image with four 5×5 kernels (25 five-bit multiplications and 25 five-bit additions per kernel) over the entire frame (16×16 pixels) at 20 kfps is equivalent to executing 1 GOPS/mW. The low power consumption per operation is several orders of magnitude smaller than a comparable system consisting of a DSP processor integrated with a CCD camera or signal-chip CNN-UM [12, 16–17].

© V. Gruev and R. Etienne-Cummings, "Implementation of steerable spatiotemporal image filters on the focal plane," *IEEE Trans. Circuits and Systems-II*, vol. 49, no. 4, pp. 233–244, Apr. 2002 (partial reprint).

5.2.5   Scalability

One of the major benefits of this architecture is scalability. Scalability is achieved by performing all computation on readout, which allows for small, scalable pixel sizes and a generic processing unit that does not depend on the size of the pixel array. The sizes of the pixel area for different technology processes are summarized in *Table 5-3*. For the 0.35 μm process, the size of the pixel is 8.75 μm × 8.75 μm, which is comparable to the size of CCD pixels. The advantage of GIP pixels (over those of a CCD) is their capability for simultaneously accessing several pixels in a given neighborhood, which enables parallel spatial and temporal convolution of the incident image. The low SNR of the GIP pixel can be improved by the addition of noise cancellation circuits, such as correlated double sampling [22]. However, the

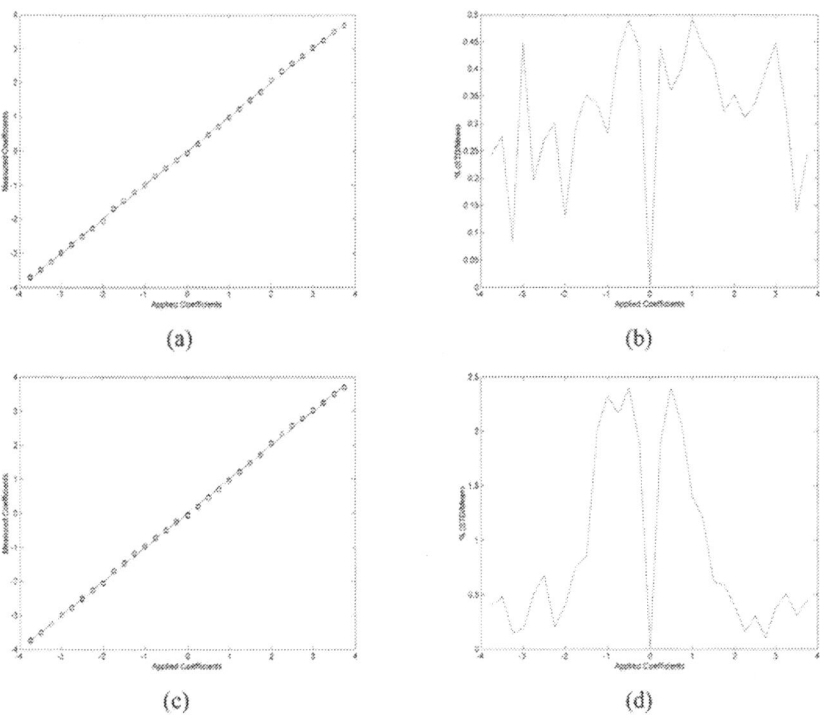

Figure 5-10. Precision of the scaling coefficients of the current in the processing units.
(a) Linear plot of the coefficients of a single processing unit for nine light intensities.
(b) Variations of coefficients within one processing unit across a range of light intensities.
(c) Linear plot of coefficients of all four processors for one light intensity. (d) Variations of coefficients across processing units for one light intensity.

optimized CCD process has better sensitivity than any of the CMOS imagers. Nonetheless, the size, light sensitivity, noise characteristics, and focal-plane processing capabilities of the GIP pixel make it ideal for image processing.

The processing unit capitalizes on the scalability of the GIP design. The area and computational speed of the processing unit is independent of the size of the photo array. For a 1 cm^2 chip fabricated in a 1.2 μm process with 333×333 photo array pixels, the ratio of the processing unit to the photo pixel array is 0.5%. For a 0.35 μm process with a 1143×1143 photo array on a 1 cm^2 die, this ratio is 0.04%. The size of the processing unit area is not affected when the size of photo array increases, leading to a high area efficiency for the processing unit. The processing unit occupies such a small percentage of the total area that the transistors in the processing unit need not be scaled at the same rate as the transistors in the pixels. This observation is crucial for maintaining or improving the precision of the processing unit. By maintaining the physical size of transistors, the precision of the processing unit can be increased due to the higher resolution and improved tolerances of technologies with smaller feature sizes. A migration to smaller feature-size technologies clearly promises improvements in both area efficiency and precision for the processing unit.

The computational speed of the processing unit is also not effected by scaling the design, because virtual ground circuits are used to mask the current bus parasitic capacitance. The virtual ground circuit keeps the bus line charged to a constant voltage at all times, which maintains the access time of the photopixel current and keeps the computational time of the convolution kernels constant. As the architecture is scaled down to smaller technologies, the total line capacitance for a 1 cm^2 die remains constant because the scaling of the oxide will be compensated by the scaling of the size of the switch transistors that connect the pixels to the bus line. The length of the bus line remains constant. Specialized scanning techniques should be considered for denser pixel arrays in order to decrease the charge-up time for the selection lines. In addition, smaller processes allow for higher operational speed; for example, the operational speed can be increased to approximately 50 MHz if the 0.35 μm process is used (see *Table 5-3*).

To maintain a constant frame rate as the size of the array is increased, the scanning frequency must be increased. Unfortunately, the pixel access circuits and processing circuits place limits on scanning speed. The importance of the virtual ground circuit in achieving fast scanning speeds under increased line capacitance has already been discussed. Current-mode

Table 5-3. Scaling properties of the GIP architecture.

	1.2 μm process	0.5 μm process	0.35 μm process
Pixel size	30 μm × 30 μm	12.5 μm × 12.5 μm	8.75 μm × 8.75 μm
Photosensitive area	13.6 μm × 13.6 μm	5.5 μm × 5.5 μm	3.85 μm × 3.85 μm
Processing unit area	1680 μm × 300 μm	700 μm × 125 μm	490 μm × 87.5 μm
Processing unit area per 1 cm^2 die (%)	0.5%	0.09%	0.04%
No. of pixels per 1 cm^2 die	333 × 333 pixels	800 × 800 pixels	1142 × 1142 pixels
Operational speed	15 MHz	30 MHz	50 MHz

processing is also important for maximizing processing speed. Virtual grounding is not being used on all GIP current summing nodes at this time, but this would be necessary for a large array scanned at high speeds. A common technique to further increase the frame rate is to divide the array into multiple parallel sub-arrays that are scanned and processed in parallel. In this case, however, the benefits of uniformity due to a single processing unit would be lost. It is clear that the GIP approach does not scale in computational complexity unless local analog memories are introduced in each pixel to store partial results for algorithmic processing. This is the major benefit of the CNN-UM over the GIP. In the future, however, some ideas may be borrowed from the CNN-UM to improve the algorithmic processing capabilities of the GIP.

5.2.6   Further applications

The spatial and temporal processing capabilities, the reprogrammability, and the parallel convolution computation of the GIP allow it to be used as a front end for many computer vision applications. The first layer of computation in these applications typically involves computing various spatiotemporal kernels in parallel. Computing these kernels at the focal plane frees up computational resources and improves the performance of these algorithms. For example, computing spatial corners requires first computing the horizontal and vertical edges of an image and then finding the points where both of these edges exist. If the temporal component for a given spatial corner is considered, spatial corners varying over time can be highlighted [23]. Computing horizontal and vertical edges as well as temporal frame differences in parallel at the focal plane greatly reduces the computational time for tracking spatiotemporal corners. Its ability to track spatiotemporal corners coupled with its low power consumption makes the GIP useful in a variety of robotics applications, such as target tracking, autonomous navigation, and obstacle avoidance [24].

Many orientation detection algorithms require the pre-convolution of imagers with spatial derivative kernels [25]. Computing simultaneous horizontal, vertical, and diagonal edges and using a weighted average can provide high-resolution information on the orientation of edge-forming objects. Combing temporal information with the edge orientation of an object leads to efficient methods for computing the direction of motion [23].

The GIP chip can compute specialized wavelet transforms as well as approximations of the discrete cosine transform (DCT) [26–27]. Computing wavelet transforms or the DCT requires several different patterns to be loaded sequentially in the scanning registers; the output of each scanning pattern is stored in an external memory (or in internal memory if available). The transforms are then completed by summing appropriately scaled values from the external memory. Implementation of these algorithms will be described in a future paper dedicated to applications of the GIP.

© V. Gruev and R. Etienne-Cummings, "Implementation of steerable spatiotemporal image filters on the focal plane," *IEEE Trans. Circuits and Systems-II*, vol. 49, no. 4, pp. 233–244, Apr. 2002 (partial reprint).

5.3 Voltage-domain image processing: the temporal difference imager

5.3.1 System overview

The temporal difference imager (TDI) consists of four main components: a photopixel array of 189 rows by 182 columns, two vertical and five horizontal scanning registers, a control-timing unit, and three difference double sampling (DDS) units. Each pixel has two outputs: a current frame output and a previous frame output. The two intensity images are presented in parallel to two independent DDS circuits where reset voltage mismatches, kTC noise, charge injection due to switching, $1/f$ noise, and fixed-pattern noise (FPN) are suppressed. The difference between the two corrected intensity images is computed in a third DDS circuit and presented outside the chip. The control-timing unit synchronizes the timing between all scanning registers and manages an efficient pipeline mechanism for computing the difference between the two consecutive images. This unit also controls the integration time of the two frames, the time between two consecutive frames, the sample-and-hold timing, and the computation timing of the DDS circuits. Different readout techniques can be executed by changing bit patterns in the scanning registers and reprogramming the control-timing unit. Hence, a fair comparison between standard readout techniques and our proposed techniques can be made on this imager.

5.3.2 Hardware implementation

The active pixel sensor (APS) cell shown in *Figure 5-11* is composed of a photodiode, two storage elements C1 and C2, switching transistors M2–M7, and readout transistors M8–M11. (Section 4.2 in Chapter 4 provides more details on APS.) PMOS transistor M1 is used to control the operation modes of the photodiode. This transistor increases the output voltage swing of the pixel by allowing the reset voltage level of the photodiode to be V_{dd}. In addition, image lag due to incomplete reset (which is evident when an NMOS reset transistor is used) is eliminated by using a PMOS reset transistor [28]. The increased output voltage swing comes at the expense of a larger pixel area, but using mainly PMOS transistors in the pixel (the exceptions are the output source follower transistors) minimizes this effect. The NMOS source follower transistors (M8 and M10) ensure that the

© V. Gruev and R. Etienne-Cummings, "A pipelined temporal difference imager," *IEEE J. Solid State Circuits*, vol. 39, no. 3, Mar. 2004 (partial reprint).

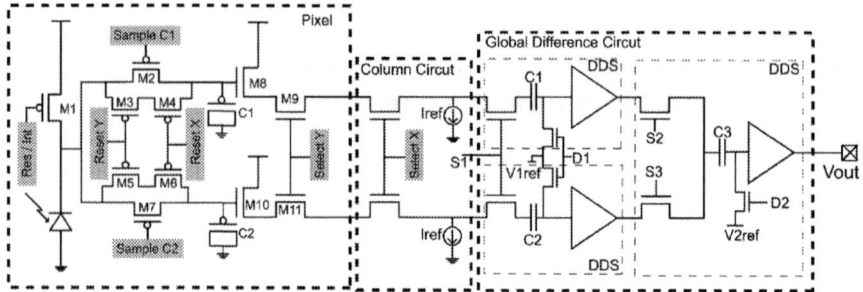

Figure 5-11. Readout circuit for the temporal difference imager (TDI).

voltage swing is between $V_{dd} - V_{th} - V(I_{bias})$ of this transistor and the minimum voltage required by the bias transistor for the follower to remain in saturation. In a traditional APS, the maximum output is described by equation (5.6).

$$V_{out\,max} = V_{dd} - V\left(I_{photodecay}\right) - V_{th,M8} - V\left(I_{bias}\right) \qquad (5.6)$$

Hence, the voltage swing is maximized at the expense of larger pixel size, and the design can be safely used at lower supply voltage levels.

The two sample-and-hold circuits are composed of a sampling transistor (M2 or M7) and a capacitor (C1 or C2), and are implemented with a transistor by connecting the source and drain to V_{ss} and the bulk to V_{dd}. Hence, the effective storage capacitance is the parallel combination of C_{gb}, C_{gd} and C_{gs}, where the subscripts g, d, s, and b indicate the transistor gate, drain, source, and bulk, respectively; this yields the maximum possible storage capacitance for the given transistor. The extra storage capacitance in the pixel will linearly reduce the photoconversion rate, but the *kTC* noise is improved only as the square root. Hence, the overall SNR will be reduced.

The sampling of the photovoltage alternates between the two sample-and-hold circuits during consecutive frames. Once the stored charges in C1 and C2 are read out to the DDS circuits, the pixel is reset and transistors M3 through M6 are turned on to discharge the holding capacitors C1 and C2. Transistors M3 through M6 allow for individual pixels to be reset, instead of the row-wise pixel reset that is common in standard APS. The reset voltage is subtracted from the integrated values in two independent DDS circuits, eliminating the voltage offset variations due to the output source follower. This technique, known as difference double sampling, improves the noise characteristics of the image. The use of two independent DDS circuits for

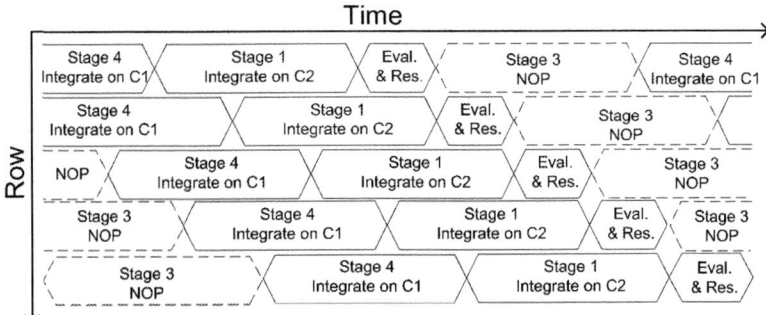

Figure 5-12. Pipeline timing.

the entire imager further improves the precision and accuracy by eliminating row FPN, which must not be present if row- or column-parallel DDS is used. After the images have been corrected, the two frames are subtracted in a third DDS circuit and the difference is provided outside the chip, together with the two intensity images.

5.3.3 Pipeline readout technique

The control-timing unit is crucial for synchronizing the different events executed in the imager. This unit controls the four-stage pipeline mechanism implemented on the chip. The timing diagram of the four stages is presented in *Figure 5-12*. The horizontal axis presents the timing events in one frame, whereas the vertical axis represents different pixels across the imager. Hence, four different tasks (stages) can be performed at the same time across different parts of the image plane. In stage 1 of the pipeline, the photovoltage is integrated and sampled just before the beginning of stage 2 on capacitor C2. Increasing or decreasing the number of integration columns in stage 1 can vary the integration period. In stage 2 of the pipeline (the pipeline consists of a single row), the previously stored photovoltage on C1, the newly integrated photovoltage on C2, and the reset values are all read out to the DDS circuit. The difference C2 – C1 is evaluated after the subtraction of the reset offset from the stored values.

The stored photovoltage on C1 is held for the entire integration period of C2. Due to the leakage currents at the holding node, this stored value will be less than the original value. The integrated photovoltage sampled on C2 will not be degraded because the difference of the two values is evaluated just as the integration period of C2 is completed. Therefore, the minimum difference between these two values will be the magnitude of the decay of C1, which will be the same for all pixels in the imager. This is very important for obtaining good precision in the difference image: an offset in the difference computation can be corrected across the entire image in the final result.

In stage 3 of the pipeline, no operations (NOPs) are executed. The length of this stage can vary between zero and the scanning time of one entire frame. When the integration time of stages 1 and 4 are equal to the time required to scan half of the columns in the imager, stage 3 does not exist. When the integration period of stages 1 and 4 are one column time, the NOP stage will be close to the scanning time of an entire frame. This stage adds some flexibility in overlapping two integration processes on the entire imager at the same time, while still controlling the integration time of each frame independently of each other. The integration times of stages 1 and 4 can be the same or different, depending of the application requirement. In most cases, the integration times of these stages are equal.

In stage 4, the photovoltage is integrated and sampled on capacitor C1. Increasing or decreasing the number of integration columns in the fourth stage can vary the integration period. Once the integration period is complete, the integrated value is stored and held in C1. The holding time of the C1 value depends only on the integration time of stage 1 and is the same for all pixels across the imager. Stages 1 and 4 of the pipeline cannot overlap, limiting the maximum integration time of both stages to half of the scanning time of one image frame. The pipeline readout mechanism allows for continuous difference evaluation as each consecutive frame is read out.

This pipeline mechanism improves the precision of the evaluated difference. To a first-order approximation, the precision of the difference strongly depends on the leakage currents at the holding nodes of C1 and C2. These leakage currents are functions of two factors: holding time and illumination. To weaken the dependency on illumination, various layout techniques were applied; these included using metal shielding methods and designing a symmetric layout to ensure that leakage currents on both holding nodes were equal. The current pipeline readout mechanism cannot eliminate the problem of leakage due to illumination. On the other hand, it reduces the problem of time dependency of the leakage currents by ensuring that the time delay between the two images is equal for all pixels. Hence, each pixel integrates for equal amount of time, and the holding time for every pixel is also equal. Since the holding times of the charges in C1 and C2 are equal and the leakage currents for all pixels can be approximated to be equal, then the offset in the temporal difference can be canceled in the final

© V. Gruev and R. Etienne-Cummings, "A pipelined temporal difference imager," *IEEE J. Solid State Circuits*, vol. 39, no. 3, Mar. 2004 (partial reprint).

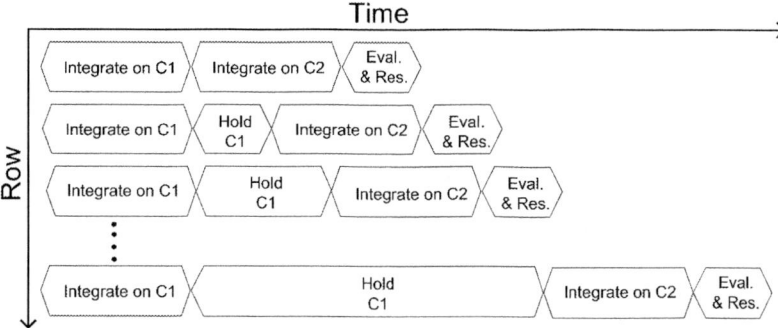

Figure 5-13. Snap shot and evaluate mode.

computation. Assuming that the two stored values in capacitor C1 and C2 are the same (V_C), the offset in the temporal difference across every pixel is described by equation (5.7).

The maximum voltage offset for these pixels is

$$V_{C1} - V_{C2} = (V_C - \Delta V_{\text{leakage}}) - V_C = \Delta V_{\text{leakage}} = \text{NOP}(I_{\text{leakage}}/C_{\text{storage}}) \quad (5.7)$$

In equation (5.7), V_{C1} is the voltage stored on capacitor C1, V_{C2} is the voltage stored on capacitor C2, $\Delta V_{\text{leakage}}$ is the decay of the value stored in capacitor C1, I_{leakage} is the reverse diode leakage current, and NOP is the holding time of the stored charge in capacitor C1 or C2. Equation (5.7) states that the voltage offset error across the entire imager is independent of the pixel position; hence, the error will be an equal offset across the entire imager.

5.3.4 Snap shot and evaluate mode

The TDI can also operate in a snap shot and evaluate mode. This mode of operation is shown in *Figure 5-13*. In this mode, the photovoltage is first integrated on capacitor C1. Then a new photovoltage is integrated on capacitor C2. After the second integration is completed, the difference between the two stored values is evaluated. Since the difference evaluation is computed in a sequential manner, the holding interval of C1 will increase as the image is scanned out. When the first pixel is evaluated, capacitor C1 has decayed by the integration time of C2. For each additional evaluation of the difference, an additional hold time is introduced. The last pixel will have the maximum hold time described by equation (5.8):

© V. Gruev and R. Etienne-Cummings, "A pipelined temporal difference imager," *IEEE J. Solid State Circuits*, vol. 39, no. 3, Mar. 2004 (partial reprint).

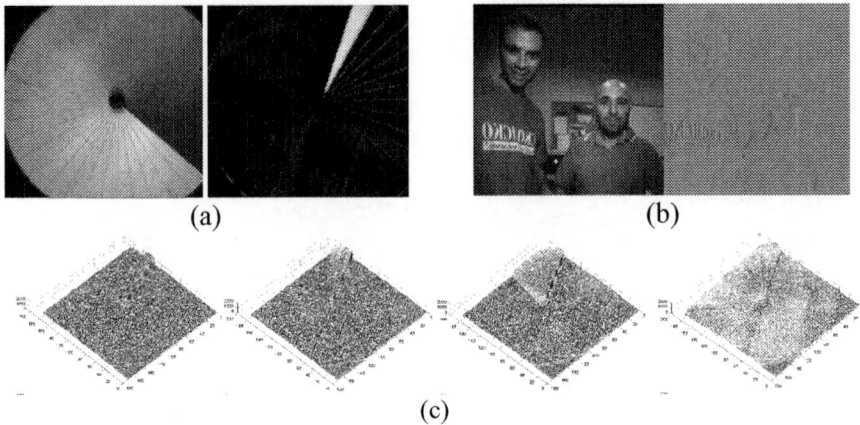

(a)   (b)

(c)

Figure 5-14. (a) Sampled intensity (left) and difference (right) images of a rotating grayscale wheel obtained with the TDI. *(b)* Sampled intensity (left) and difference (right) images obtained with the TDI. *(c)* Temporal difference images for the grayscale wheel at increasing (left to right) rotation speeds.

$$(\Delta t)_{\max} = t_{(\text{int of C2})} + MN t_{\text{clk}} \qquad (5.8)$$

In equation (5.8), $t_{(\text{int of C2})}$ is equal to the integration time of C2. M and N are the dimensions of the imaging array. The additional hold time, which increases for each scanned pixel, introduces a different offset error for each pixel. If the light intensity dependency of this offset error is ignored, the offset should be linearly increasing across the entire imager. This readout technique will require offline calibration and correction, and may not be suitable for applications requiring direct and fast computations of temporal difference [28].

5.3.5   Results

Real-life images from the TDI are shown in *Figures 5-14(a)* and *(b)*. In each figure, the intensity image is on the left side and the absolute difference is on the right side. The contour of the rotating wheel in *Figure 5-14(a)* is clearly visible in the corresponding difference image. *Figure 5-14(c)* shows the temporal differences of the grayscale wheel at different rotational speeds. The grayscale wheel contains 36 different grayscale values, with 10 degrees of spatial distribution for each value and constant grayscale increments. In the first image on the left in *Figure 5-14(c)*, the grayscale wheel is rotated

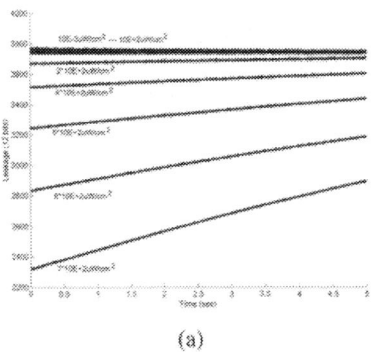 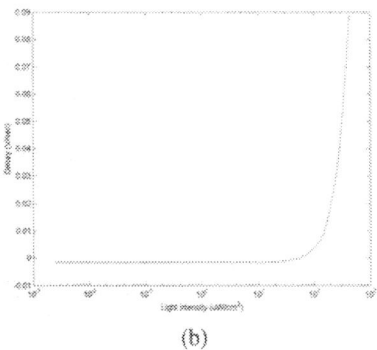

(a)   (b)

Figure 5-15. (a) Leakage currents at selected light intensities. *(b)* Decay of stored value as a function of light intensity.

slowly and the temporal difference image records only a small difference between the two consecutive images. Due to the abrupt difference between the first and last grayscale values on the grayscale wheel, the temporal derivative computes a high difference in this border region. The temporal difference in the rest of the image is constant and low due to the constant grayscale increments. As the rotational speed of the grayscale wheel is increased, the temporal difference image shows greater difference (overlap) between two consecutive images. In the last case, when the rotational speed is the highest, a wide region of high temporal difference values is recorded. The temporal difference in the rest of the image also has higher values compared to the other cases where the wheel was rotated at slower speeds. This is due to the increased overlap between the two consecutive images, which leads to higher temporal difference values.

Figures 5-15(a) and *5-15(b)* demonstrate the leakage at the holding nodes as a function of light intensity. *Figure 5-15(a)* presents discharge curves at the holding node C1 for several different light intensities. For light at low intensities (10^{-2} μW/cm^2) and mid-level intensities (10^2 μW/cm^2), the slopes of the discharge curves are negligible, corresponding to less than 10 mV/sec decay (*Figure 5-15(b)*). This decay rate allows for frame rates as slow as 3 fps with a temporal difference precision of 8 bits. For very high illumination intensities, the slope of the discharge currents increases to about 100 mV/sec. The parasitic reverse diodes at the holding node are in deep reverse bias, which results in high leakage currents.

When operating the TDI at 30 fps with a decay rate of 100 mV/sec, a temporal difference can be computed with 8-bit accuracy (the output swing of the pixel is 3 V; a 12-bit low-noise ADC was used to digitize the image).

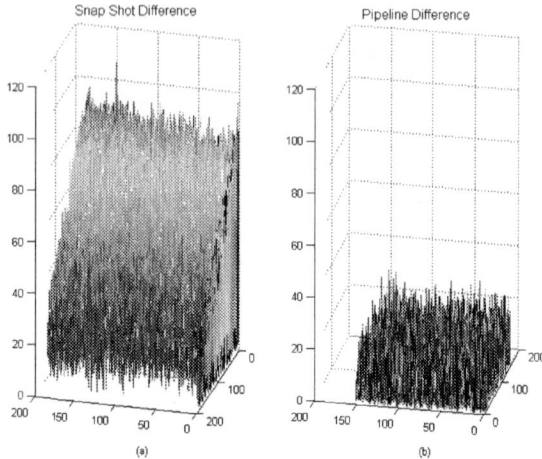

Figure 5-16. Comparison between *(a)* snap shot and *(b)* pipelined modes of operation at 150 mW light intensity.

Equation (5.9) indicates the bits of precision as a function of decay rate, frame rate, and voltage swing of the pixel:

$$error\,(\%) = \frac{decay\ rate}{fps \times voltage\ swing} \times 100 \tag{5.9}$$

The leakage currents at the holding nodes limit the precision of the temporal difference. If a snap-shot-and-evaluate readout technique is used, this precision will be further degraded. Using the pipeline readout technique helps to eliminate this problem, however, as shown in *Figure 5-16*. This figure compares the operation of the TDI in the two different readout modes.

The snap shot and evaluate mode is the mode usually discussed in the literature [29–30]. In this mode, two consecutive snap shots are obtained and their difference is computed. Unfortunately, the leakage currents strongly influence the accuracy of the computed difference. The first pixel evaluated will have the least holding time and leakage. As the rest of the image is evaluated, each additional pixel will have additional time and leakage. The last pixel evaluated in the image will have the longest time delay. The large leakage currents in the last pixels will greatly affect the accuracy of their evaluated differences. The accuracy of these differences across the imager is demonstrated in *Figure 5-16(a)*. The slope is evident in both the *x* and *y* directions as the image is scanned and as the differences are evaluated.

In the second mode of operation, the TDI takes advantage of the pipeline architecture. Since all pixels have the same holding time when operating in this mode, the variation due to leakage currents of the difference is minimized. The results of this can be seen in *Figure 5-16(b)*. The mean difference in this case is 16.5 mV with 0.14% variations of the maximum value. The advantages of the pipeline architecture compared to the snap shot operational mode are readily apparent.

The total power consumption is 30 mW (with three DDS circuits each consuming 9 mW of power) at 50 fps with fixed-pattern noise at 0.6% of the saturation level, which makes this imager attractive for many applications. The greater than 8-bit precision for the difference between two consecutive images and the relatively low FPN are the major advantages of this architecture.

5.4   Mixed-mode image processing:
         the centroid-tracking imager

5.4.1   System overview

This imager consists of two subsystems, the APS imager and the centroid tracker. The APS array obtains real-time images of the scene and the centroid tracker computes the location of moving targets within the scene. Each can be operated independently of the other. In this design, no resources are shared between the two except for the focal plane itself.

Figure 5-17 shows the floor plan of the array and edge circuitry. To facilitate tiling, the pixel for centroid computation is exactly twice as long on each side as the APS. Pixels in the same row are of the same type, and the array rows alternate between centroid-localization pixels and APS. Due to the difference in size of the pixels, APS rows are 120 pixels across whereas each centroid row contains 60 pixels. The non-uniform arrangement of pixels was motivated by a desire to increase APS resolution for better image quality. Unfortunately, this decision was directly responsible for APS matching and performance problems, which will be discussed in more detail below. Digital lines are run along rows to keep the digital switching transients for one style of pixels from coupling to signals in the other. The chip was fabricated in a standard analog 0.5 µm 1P3M CMOS process.

5.4.2   APS imaging subsystem

The design of the APS pixel follows the same basic three-transistor, one-photodiode design pioneered by Fossum et al. [31] (shown in *Figure 5-18*). Included are a reset transistor, an output transistor, and a transistor select switch to address the pixel. The structure is simple, with no provision for electronic shuttering. It is optimized primarily for density and secondarily for fill factor. All transistors in the pixel are NMOS to reduce the area of the pixel. (More details on APS are presented in Chapter 4, section 4.2.)

The row circuitry is composed of two cyclical shift register chains: one for row reset signals and the other for row select signals. Each row of pixels receives reset and row select signals from one stage of each chain. Clocking these shift registers advances their bit pattern forward by one row and starts the readout of the next row. The reset shift register can be preloaded with blocks of ones and zeros in a flexible way, allowing the integration time for

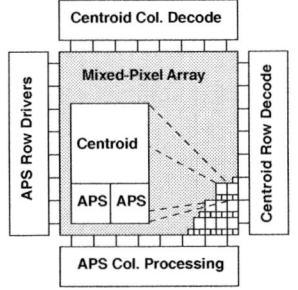

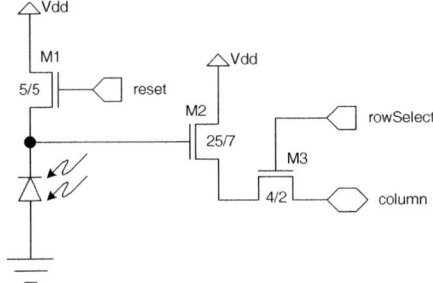

Figure 5-17. System-level view of the centroid-tracking imager chip.

Figure 5-18. APS pixel schematic.

each row to be specified as a modifiable fraction of the total frame time. This can be viewed as a "rolling shutter." In addition, there is circuitry on the reset lines to facilitate reset timing on a shorter time scale than one row clock. A separate global signal, labeled "directReset" in *Figure 5-20,* is ANDed with the signal of each row from the shift register. Using this signal, integration can be stopped and reset initiated in the middle of the output cycle of a single row. This is especially important for the operation of the correlated double sampling (CDS) system described below. It also permits easy prototyping, allowing a simple method for examining the operation of the pixels in a single row without running the entire array.

Each column of pixels has its own dedicated processing circuitry. The column circuitry starts with its most important block, the CDS circuit [32–33]. This circuit subtracts the reset voltage from the signal voltage, ensuring that only the difference between the two is measured and not the absolute signal level itself. This drastically reduces offset errors in readout. It also compensates for noise and different reset voltage levels resulting from different light intensities during reset. The CDS is implemented as a simple switched capacitor circuit, producing a single-ended output voltage. A fully differential circuit would have exhibited more immunity to power supply ripple and interference from other signals, but these factors are not compelling in this application. Therefore, a simpler method is used to shorten design time and to minimize the area of the layout. Efficient use of space is especially important for this circuit because it is used in each column. A simple 7-transistor (diffamp and inverter) opamp with Miller compensation in a unity-gain configuration follows the CDS circuit for output buffering. The end of the column circuit employs yet another shift register chain to sequentially activate the switches that output one column voltage at a time to the single-pin output. *Figure 5-19* shows a schematic of the CDS circuit.

© M. Clapp and R. Etienne-Cummings, "Dual pixel array for imaging, motion detection and centroid tracking," *IEEE Sensors Journal*, vol. 2, no. 6, pp. 529–548, Dec. 2002 (partial reprint).

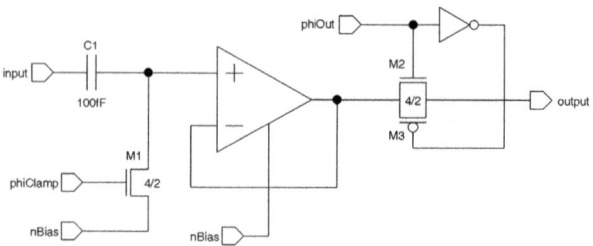

Figure 5-19. CDS, column buffer, and output switching circuit.

5.4.3   Centroid-tracking subsystem

The basic function of this subsystem is to compute the centroid of all the pixels that have brightness levels varying with time. This approximates finding the centroid of a moving object. A moving object will at least cause pixels to change at its edges (in the case of a solid-colored object, for example) and usually many pixels within the image of the object will also change (due to details or texture). In either case, the centroid of all time-varying pixels will be close to the center of the object. This scheme works most accurately for small objects because no points on a small object are very far from the centroid. The uncertainty in pixel activity detection will thus cause a smaller possible error in the computation of centroid position. In this particular design, implementation details necessitated that only an increase in brightness is detected; the reasons for this modification are explained below. With this alteration, moving bright objects on a dark background should theoretically be tracked by their leading edge, and dark objects on a bright background by their trailing edge. This may cause additional deviation from the true centroid in some situations. However, most real-world objects have visible texture and are not solid-colored. In these situations, many pixels inside the outline of the object will be activated in addition to the outline pixels, lessening the impact of ignoring intensity decreases. The output of this subsystem is a set of two voltages: one for the *x* position and one for the *y* position.

The method employed to detect pixel brightness changes can be considered as a simplified form of an address event representation imager [34–36]. The only output from each activated pixel is the row and column of that pixel. Edge circuitry then processes the activated rows and columns to find the centroid. Moving the more complicated processing to the edges of the array keeps the pixel size smaller and helps to increase the fill factor for the motion-sensitive pixels.

5.4.4   Centroid pixel

The pixel itself starts with a photodiode, which is biased by an NMOS transistor with its gate voltage fixed (see *Figure 5-20.*) The voltage at the source of this load NMOSFET will be proportional to either the logarithm or the square root of the incident light intensity, depending on whether the photodiode current operates the NMOS in the subthreshold region or the above-threshold region, respectively. Since the goal is to detect a relative change in brightness, the circuit is designed to be sensitive to the same multiplicative change in photocurrent at any absolute brightness level. Such contrast-sensitive photodetectors have also been observed in biological visual systems [37]. The logarithmic transfer function of the subthreshold transistor translates a multiplicative increase or decrease in photocurrent into an additive increase or decrease in output voltage, simplifying the task for the next stage of the pixel. The square root function does not share this property exactly, but it has a similar curve and approximates a logarithm. Fortunately, the pixels of the centroid-tracking imager chip operate in the subthreshold region for most light levels. Light intensities of over $10 \, \text{mW/cm}^2$ are required to generate over 1 nA of photocurrent, and in practice even extremely bright light conditions do not exceed $1 \, \text{mW/cm}^2$ at the photosensor. The photosensitive voltage is AC-coupled to the rest of the pixel through a PMOS capacitor with the well tied to the drain and source. The rest of the pixel consists of a resettable comparator circuit (implemented using a biased CMOS inverter) and a feedback switch. The inverter includes a cascode transistor to enhance gain.

Operation of the pixel starts with the reset of the comparator block within the pixel. The inverter feedback switch is closed, input is made equal to output, and the inverter settles at its switching voltage. At this time, the difference between the photodiode cathode and inverter input voltages is stored across the PMOS capacitor. The PMOS capacitor is held in inversion, since the inverter reset voltage is significantly lower than the photodiode voltage. When the switch is opened, the inverter goes into open-loop operation. As the light level on the photodiode increases, the voltage on its cathode will decrease. Since the input to the inverter circuit is floating (high impedance), its voltage will now track the voltage on the photodiode, offset by the voltage across the capacitor itself. When the voltage decreases by a given amount ΔV corresponding to a given factor increase in photocurrent, the inverter will trip and its output will go high. If light on the pixel decreases, however, no event will be signaled because the inverter will move even farther away from its switching threshold.

© M. Clapp and R. Etienne-Cummings, "Dual pixel array for imaging, motion detection and centroid tracking," *IEEE Sensors Journal*, vol. 2, no. 6, pp. 529–548, Dec. 2002 (partial reprint).

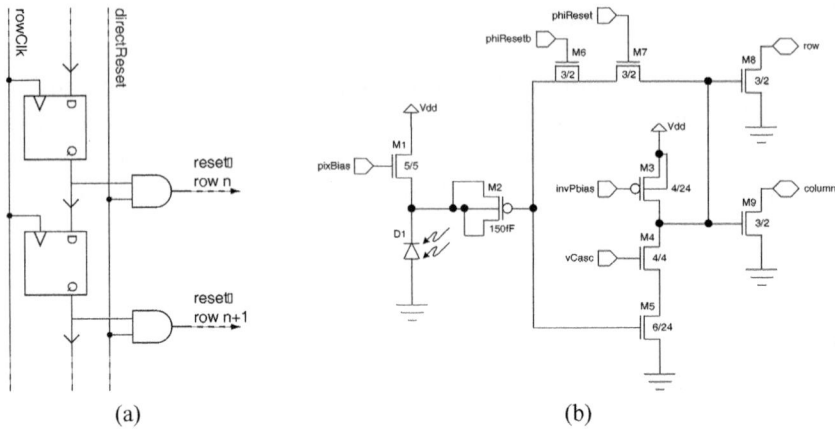

Figure 5-20. (a) APS row circuitry. *(b)* Schematic of the centroid-tracking pixel.

The amount of change in incident light necessary to trigger the pixel after reset is released depends on the setting of V_{pbias} and the intensity of light incident during reset. The setting of V_{pbias} will set the reset voltage of the inverter. To maximize the gain of the inverter and to save power, subthreshold-level currents are used in the inverter. Equation (5.10) follows from equating the drain currents of the PMOS and NMOS transistors:

$$I_{0N}e^{\frac{V_{out}\kappa_n}{U_t}} = I_{DN} = I_{DP} = I_{0P}e^{\frac{(V_{dd}-V_{pbias})\kappa_P}{U_t}} \tag{5.10}$$

where U_t is the thermal voltage, κ_N and κ_P are subthreshold slope factors, and I_{0N} and I_{0P} are process-dependent and gate-geometry-dependent factors relating fundamental currents of the subthreshold transistors. The equation for output reset voltage is thus

$$V_{\text{invreset}} = \frac{U_t}{\kappa_N}\ln\left(\frac{I_{0P}}{I_{0N}}\right) + \frac{\kappa_P}{\kappa_N}\left(V_{dd} - V_{\text{pbias}}\right) \tag{5.11}$$

Since I_{0P} is significantly less than I_{0N}, it can be seen from this equation that $V_{\text{invreset}} < V_{dd} - V_{\text{pbias}}$. The difference between V_{invreset} and V_T of the NMOS row and column pull-down transistors determines the initial output ΔV necessary to cause the NMOS pull-down transistors to conduct and signal a change:

© M. Clapp and R. Etienne-Cummings, "Dual pixel array for imaging, motion detection and centroid tracking," *IEEE Sensors Journal*, vol. 2, no. 6, pp. 529–548, Dec. 2002 (partial reprint).

$$\Delta V_{out} = V_{TN} - V_{invreset} \qquad (5.12)$$

Dividing this ΔV_{out} by the gain of the inverter yields the ΔV_{in} necessary to sufficiently move the output:

$$\Delta V_{in} = \frac{\Delta V_{out}}{A_{inv}} \qquad (5.13)$$

where the gain of the inverter with subthreshold drain current is

$$A_{inv} = \frac{-g_{m5}}{\left(\dfrac{g_{ds4}}{g_{m4}}\right)g_{ds5} + g_{ds3}} \qquad (5.14)$$

This becomes

$$A_{inv} = \left(\frac{-\kappa_N}{U_t}\right)\left(\frac{1}{\dfrac{U_t}{V_{ON4}\kappa_N}\dfrac{1}{V_{ON5}} + \dfrac{1}{V_{OP3}}}\right) \approx \frac{-\kappa_N V_{OP3}}{U_t} \qquad (5.15)$$

yielding

$$\Delta V_{in} = -\frac{\Delta V_{out} U_t}{\kappa V_{OP}} \qquad (5.16)$$

or

$$\Delta V_{in} = -\frac{\left(V_{TN} - V_{invreset}\right) U_t}{\kappa V_{OP}} \qquad (5.17)$$

The terms V_{ON} and V_{OP} are the early voltages of the NMOS and PMOS transistors in subthreshold states. Because of the cascode transistor, the g_{ds} of transistor M5 no longer makes a significant contribution to the gain, and can be neglected in the final expression. Note that ΔV_{in} is negative due to the

© M. Clapp and R. Etienne-Cummings, "Dual pixel array for imaging, motion detection and centroid tracking," *IEEE Sensors Journal*, vol. 2, no. 6, pp. 529–548, Dec. 2002 (partial reprint).

negative gain of the inverter. This equation describes the basic operation of the pixel. More details, including the effect of the coupling of the switching voltages, will be presented in section 5.4.6 and later sections.

The inverter drives two NMOS pull-down transistors that are attached to the particular row and column lines associated with the pixel. These lines are set up in a wired-OR configuration, with weak PMOS pull-up transistors on the edges of the array. Switches can disconnect the array from the edge circuitry to avoid current draw during reset.

5.4.5   Centroid edge circuitry

The centroids of the activated rows and activated columns are computed separately to arrive at a final (x, y) coordinate for the two-dimensional centroid. A center-of-mass algorithm is employed, resulting in sub-pixel precision.

The following is a description of the edge circuitry operation specifically for the column edge; row edge operation works identically. *Figure 5-21* shows the schematic of this module. The edge of the centroid subsystem receives a series of column outputs corresponding to each column of the centroid pixel array. Columns containing pixels that have experienced an increase in their brightness will show up as a logic low. The center-of-mass calculation computes a weighted average of every activated column using the column position as weight. For example, if only columns 20 and 21 have been activated, the result of the center-of-mass calculation would be 20.5. This example also illustrates sub-column position precision. The position weights are represented as a set of voltages from a resistive ladder voltage divider with as many taps as there are columns. These voltages are buffered using simple five-transistor differential amplifiers. A column with a low (activated) output will first set an SR flip-flop, locking it high until the flip-flop is reset with an externally provided reset signal. The outputs of the SR flip-flops turn on weak PMOS transistors operating in the ohmic region, which connect the column weight voltages to the centroid output node. The PMOS transistors have a width/length ratio of 4/22, and are turned on by lowering their gates fully to ground. The weight voltages of all active columns are then connected to the common node through the PMOS pseudo-resistors, and this network of voltages interconnected through identical pseudo-resistors computes the average of all voltages connected. The output voltage is thus the center of mass value of all active columns.

In the ideal case, all activated PMOS resistors would be in the linear region so that V_{ds} has a linear relation to the current flowing, approximating

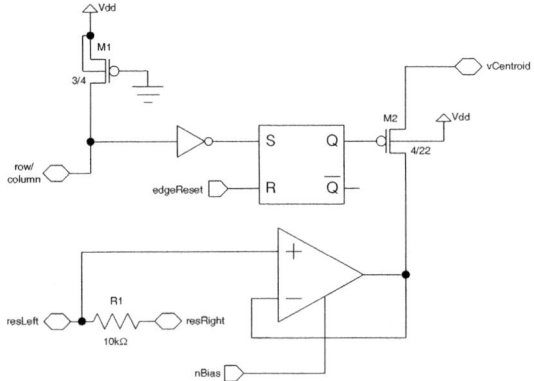

Figure 5-21. One segment of the centroid edge circuitry.

a true resistor. For a PMOS to be operating in the linear region, the condition $-V_{ds} < -V_{gs} + V_T$ must hold. Equivalently, $V_{ds} > V_{gs} - V_T$. It must be true that $V_{gs} < 0 - V_{laddermin}$, where $V_{laddermin}$ is the low voltage of the resistive ladder. Therefore, the sufficient condition for linear operation can be expressed as

$$V_{ds} > -V_{laddermin} - V_T(V_s) \qquad (5.18)$$

The threshold voltage is dependent on the source voltage due to the bulk effect. In addition, $V_{ds} > V_{laddermin} - V_{laddermax}$ must always be true, because the minimum drain voltage possible is the minimum voltage of the resistive ladder and the maximum is $V_{laddermax}$, the maximum value of the voltage ladder. The values of $V_{laddermin}$ that satisfy the inequality

$$V_{laddermin} - V_{laddermax} > -V_{laddermin} - V_T \qquad (5.19)$$

or

$$V_{laddermin} > \frac{V_{laddermax} - V_T}{2} \qquad (5.20)$$

will cause V_{ds} to absolutely satisfy the condition for operating in the linear region. For a typical V_{TP} of -1 V and $V_{laddermax}$ of 2.8 V, the low voltage of the resistive ladder must therefore be 1.95 V to guarantee that all PMOS transistors will operate in the linear region.

© M. Clapp and R. Etienne-Cummings, "Dual pixel array for imaging, motion detection and centroid tracking," *IEEE Sensors Journal*, vol. 2, no. 6, pp. 529–548, Dec. 2002 (partial reprint).

In the normal mode of operation, the low voltage of the resistive ladder used in the centroid-tracking imager chip is 1.8 V, and the high ladder voltage is 2.8 V. In the worst case, it is possible that a PMOS transistor sometimes will not be operating in the linear region, and hence will dominate the averaging operation due to its higher conductance. In practice, however, moving objects are localized. As long as there is only a single moving object in the scene, the activated rows and columns will be in close proximity to one another. Hence, the V_{ds} voltages between the reference voltages and their average will stay small enough to keep each PMOS operating in the linear region.

It should be noted that each pixel position factors into the center of mass calculation with equal weight. Because the region of interest is defined as everywhere that pixel light intensity has changed, it is necessary to assume that every point has a weight of 0 or 1. It is possible to imagine other functions, such as one that weights each pixel by how much its light intensity has changed. However, it is unclear whether this is a desirable metric. Therefore, it is assumed for this system that the only meaningful algorithm is a binary condition: change or no change.

In addition, this circuit does not consider the number of pixels activated in a column or row. It gives every column or row the same weight independent of the number of activated pixels. Instead of noting the actual centroid of the pixels that are activated, it detects the centroid of a rectangular box coincident with the edges of the region of activated pixels. This was chiefly an implementation-related optimization. It is much easier for the edge circuitry to note activity/non-activity than to include how much activity a certain row or column contains. For most objects, the centroid of a coincident rectangular box is a good approximation of their true centroid. The main drawback of this modified centroid is that single outlying pixels are given as much weight as those that are clustered together. Thus, false activity registering on one pixel gives the row and column of that pixel the same weight in centroid calculations as the rows and columns that contain many pixels responding to the image of the real object. This is a regrettable disadvantage, but it can be justified by the much-simplified implementation of the current scheme.

5.4.6   APS analysis

The APS subsystem is composed of two main units: the APS pixel and the edge circuitry that both controls the pixels and helps to process output analog data.

The APS pixel used here is a well-proven 3T design (reset, amplifier, and readout switch transistors), which has been well analyzed in other papers [16, 38]. The main characteristics of this specific implementation are summarized below. *Figure 5-18* shows a schematic of the APS pixel and *Figure 5-19* shows the column readout circuit.

5.4.7   APS pixel linearity, gain, and sensitivity

The gain of the pixel from incident light to output voltage is a function of only a few circuit elements. The first is the integrating capacitance of 94.2 fF on the photodiode node. This sets the conversion gain of the photodiode at 1.70 μV/e$^-$. Following the input capacitance is the gain of the pixel gate-source amplifier. The output impedance of the column current sources and g_{mb} set this gain at 0.77. The switched capacitor of the CDS circuit ideally performs subtraction of voltages with a gain of one. Leakage currents and coupling in the switches will introduce error, but because this is not a gain error, its gain can be assumed to be one for the purposes of this analysis. Following the switched capacitor is a two-stage opamp connected in unity-gain configuration. As such, its actual gain is more like $A/(1+A)$, where A is the gain of the opamp. For the design considered here, A is around 15,000, which makes the gain of the buffer configuration virtually unity.

At this point, the total gain of the system is 1.31 μV/e$^-$. To translate this into a useful figure, it is necessary to convert the units of e$^-$ to units of (μW/cm^2)·s by assuming a quantum efficiency of 20% and a wavelength of 600 nm, and by noting that the photodiode area is 30.87 μm^2. The resulting gain, equating voltage to light intensity and integration time, is 244 μV/((μW/cm^2)·ms).

5.4.8   APS noise

The noise of the APS system begins with the photodiode itself. Photon shot noise and dark current shot noise are described as follows:

$$\left\langle v_{photon}^2 \right\rangle = \frac{I_{photo}\Delta t_{reset}}{C_{pdiode}^2} q \tag{5.21}$$

$$\left\langle v_{dark}^2 \right\rangle = \frac{I_{dark}\Delta t_{reset}}{C_{pdiode}^2} q \tag{5.22}$$

With $I_{photo} + I_{dark} \approx 2$ pA, $\Delta t_{reset} = 926$ μs, and the capacitance of the photodiode node at 94.2 fF, the total noise from current through the photodiode comes to about 33.4×10^{-9} V^2, or 183 μV$_{rms}$.

Reset noise is calculated to be 210 μV$_{rms}$ from the following basic equation:

$$\left\langle v_{pixreset}^2 \right\rangle = \frac{kT}{C_{pdiode}} \qquad (5.23)$$

This noise figure is only appropriate for reset times that are long enough for the photodiode to reach a steady state during reset. The usual mode of operation for the centroid pixel involves a reset as long as a full-row readout time (926 μs), which is long enough for the pixel to reach steady state reset at moderately high light levels. However, for lower light levels, the non-steady-state noise energy relation should hold:

$$\left\langle v_{pixreset2}^2 \right\rangle \approx \frac{kT}{2C_{pdiode}} \qquad (5.24)$$

This corresponds to a voltage of 148 μV$_{rms}$. This thermal noise figure gives the fundamental noise floor of the images regardless of matching.

Noise is also associated with the output follower pixel amplifier and the input to the CDS circuit. During the clamping of the CDS capacitor, the noise can be modeled as kT/C noise with the CDS capacitor at 100 fF. This noise amounts to

$$\left\langle v_{cdsclamp}^2 \right\rangle = \frac{kT}{C_{CDS}} \qquad (5.25)$$

or 41.43×10^{-9} V^2.

After the output side of the CDS clamp is unclamped, the total noise power becomes the sum of the noise contributions from the pixel follower amplifier and the column bias transistor. The two noise sources present are $1/f$ and transistor shot noise. The noise in the currents of these transistors includes $\langle i_{th}^2 \rangle$, contribution of the thermal noise of each transistor, and $\langle i_f^2 \rangle$, the contribution of the $1/f$ noise for each transistor, where

© M. Clapp and R. Etienne-Cummings, "Dual pixel array for imaging, motion detection and centroid tracking," *IEEE Sensors Journal*, vol. 2, no. 6, pp. 529–548, Dec. 2002 (partial reprint).

$$\left\langle i_{th}^2 \right\rangle = \frac{8kTg_m}{3} \tag{5.26}$$

and

$$\left\langle i_f^2 \right\rangle = \frac{K_F I_b^{A_F}}{f C_{ox} L_{eff}^2} \tag{5.27}$$

The total noise current flowing in the column line is

$$\left\langle i_{col}^2 \right\rangle = \left\langle i_{thM1}^2 \right\rangle + \left\langle i_{thM2}^2 \right\rangle + \left\langle i_{fM1}^2 \right\rangle + \left\langle i_{fM2}^2 \right\rangle \tag{5.28}$$

and the resulting noise (in energy units of V^2) to the input of the non-clamped CDS circuit is

$$\left\langle v_{col}^2 \right\rangle = \frac{\left\langle i_{col}^2 \right\rangle}{g_{m1}^2} \tag{5.29}$$

The noise contribution of the follower amplifier is denoted by $\left\langle v_{amp}^2 \right\rangle$, and is calculated to be 185.54×10^{-9} V^2 for the buffer amplifiers of the centroid-tracking imager, corresponding to 431 μV_{rms} [39].

In addition to all of these fundamental noise sources [40], there are also the unwanted variations in processing in the pixels that are collectively named fixed-pattern noise (FPN). The dominant phenomenon of FPN is the random variation in the threshold voltage of the reset and pixel amplifier transistors. The CDS circuit should eliminate the effects of this variation, and should eliminate the $1/f$ noise sources in the circuit as well. However, the reset is sampled after the signal, and the two voltages are not part of the same integration cycle; thus the kT/C noise from the sampled reset is not correlated to the noise from the integration cycle. This means the kT/C noise is not eliminated, and must still be included in the total noise figure. Indeed, it must be counted twice because the reset noise power from the actual integration cycle and from the sampled reset will both add to the total noise. The noise contributions of the column *with* CDS cleanup of the $1/f$ noise has been labeled here as $\left\langle v_{colcds}^2 \right\rangle$, where

© M. Clapp and R. Etienne-Cummings, "Dual pixel array for imaging, motion detection and centroid tracking," *IEEE Sensors Journal*, vol. 2, no. 6, pp. 529–548, Dec. 2002 (partial reprint).

$$\left\langle v_{\text{colcds}}^2 \right\rangle = \frac{kT}{C_{\text{column}}} \tag{5.30}$$

Since the noise remaining after CDS contains only thermal components, it can be reduced to a kT/C term. This leads to a final noise expression, after CDS, of

$$\left\langle v_{\text{apstotal}}^2 \right\rangle = \left\langle v_{\text{photon}}^2 \right\rangle + \left\langle v_{\text{dark}}^2 \right\rangle + 2\left\langle v_{\text{pixreset2}}^2 \right\rangle + \left\langle v_{\text{cdsclamp}}^2 \right\rangle + \left\langle v_{\text{colcds}}^2 \right\rangle + \left\langle v_{\text{amp}}^2 \right\rangle$$

or

$$\left\langle v_{\text{apstotal}}^2 \right\rangle = \frac{\left(I_{\text{photo}} + I_{\text{dark}} \right) \Delta t_{\text{reset}}}{C_{\text{pdiode}}^2} q + \frac{kT}{C_{\text{pdiode}}} + \frac{kT}{C_{\text{CDS}}} + \frac{kT}{C_{\text{column}}} + \left\langle v_{\text{amp}}^2 \right\rangle \tag{5.31}$$

Adding all of these noise contributions together gives

$$\left\langle v_{\text{apstotal}}^2 \right\rangle = 304.3 \times 10^{-9} \, \text{V}^2 \tag{5.32}$$

for a final noise amplitude of 552 μV_{rms}, neglecting $1/f$ noise.

5.4.9   APS dynamic range and SNR

The output voltage range of the APS system is originally limited by the maximum reset voltage in the pixel, minus the lowest voltage for reliable photodiode operation.

In this case, reset is approximately $V_{dd} - V_{TN}$, or 3.3 V – 1.0 V = 2.3 V for an NMOS transistor (including the bulk effect). The reset transistor will still be supplying current to the photodiode during the reset cycle. Exactly how much current the photodiode draws will be determined by the light intensity falling on the pixel at the time of reset, and the final voltage of the pixel will reflect this. Since these factors can and do vary during the operation of the imager, the real reset voltage also varies. Part of the function of the CDS circuit is to compensate for this normal variance of the reset signal. The follower in the pixel causes the column voltage to drop by a further amount equal to V_{TN}, which together with the bulk effect reduces the maximum (reset) voltage on the column to 1.3 V.

© M. Clapp and R. Etienne-Cummings, "Dual pixel array for imaging, motion detection and centroid tracking," *IEEE Sensors Journal*, vol. 2, no. 6, pp. 529–548, Dec. 2002 (partial reprint).

The minimum voltage of the column is dictated by the column current sources: they must stay in saturation. For this to be true, the column voltage cannot drop to less than $V_{ds} > V_g - V_T$, or

$$V_{ds} \geq \sqrt{\frac{I_D 2}{K'_N} \frac{L}{W}}$$

(5.33)

where I_D is the saturated bias current for the column. For this chip and for the bias of 260 nA used in these columns, it can be calculated that V_{ds} must be greater than 85 mV to stay in saturation. This gives a practical minimum voltage of 100 mV. The output range is therefore about 1.2 V.

With an output range of 1.2 V and a noise level of 455 μV_{rms}, the signal to noise ratio is 68 dB.

5.4.10   APS speed

The design goal for imaging frame rate was 30 fps. The APS subsystem easily meets this specification for speed. Faster frame rates are possible, but there is a direct trade-off between exposure time and frame rate, with faster rates requiring higher light levels. The absolute limit on speed is governed by the column current sources that bias the source followers in the pixels during operation. These current sources are normally biased to around 260 nA for low-power operation. This current drive, combined with the column line capacitance of 200 fF, gives a maximum fall time of 1.3 V/μs. Therefore, the worst case settling time for one column is about 925 ns with a 1.2 V voltage range. The settling time for each CDS amp to be switched onto the pixel bus is 20 ns. Thus, the columns in a row take 925 ns to settle, and each pixel clocked out takes 20 ns to settle. In this imager with 36 rows and 120 columns, the maximum frame rate is estimated to be 8300 fps if the exposure problems associated with short integration times are ignored.

In practice, the minimum frame rate is set by the desired SNR of the imager and the light level to be imaged. The maximum integration time per frame is 35/36 of the frame duration. Hence, the formula for the frame rate is

$$r_{frame} \leq \frac{(\text{light level})(\text{light to voltage gain})}{(SNR)(\text{system noise})}$$

(5.34)

Using values computed earlier in this chapter, this becomes

$$r_{\text{frame}} = \frac{(\text{light level})\left(244\dfrac{\mu V}{(\mu W/cm^2)\cdot ms}\right)}{(SNR)(455\,\mu V)} \tag{5.35}$$

5.4.11   Correlated double sampling analysis

The CDS circuitry is essentially very simple: a capacitor, a switch, and a buffer opamp. Leakage of the floating capacitor node is the biggest potential problem to be faced. The severity of the effects of such leakage on the output signal will be estimated in this section.

Given the known dark current of the APS pixel, it is estimated that leakage from the drain diffusion of the clamping switch is 20.25 aA. It is also known that the CDS series capacitor has a value of 100 fF. These data allow the calculation of the voltage decay rate due to leakage:

$$\frac{\Delta V}{\Delta t} = \frac{20.25\,\text{aA}}{100\,\text{fF}} = 203\,\text{pV}/\mu s \tag{5.36}$$

For a 33 ms frame, the 925 μs row readout time will cause this voltage to decay by 188 nV. This figure is far below the noise level of the APS system and can safely be ignored. Only a frame rate roughly 200 times slower than this (about 1 frame every 6.6 s) would cause this leakage to be significant compared to the noise of the system.

5.4.12   APS power consumption

The power consumption of the whole subsystem is the sum of the power requirements for the digital row circuitry, the pixel reset current, the pixel amplifier output current, the CDS circuitry, and finally the digital shift registers for outputting each pixel. Assuming a normal photocurrent of 2 pA/pixel, which is observed under ordinary indoor lighting conditions, the total current of the APS subsystem is estimated to be 493 μA. The data in *Table 5-4* shows that the dominant current draw is the due to the biases of the CDS buffer amplifiers.

© M. Clapp and R. Etienne-Cummings, "Dual pixel array for imaging, motion detection and centroid tracking," *IEEE Sensors Journal*, vol. 2, no. 6, pp. 529–548, Dec. 2002 (partial reprint).

Circuit	Current Consumption (µA)
Column biases	31.2
Photocurrent (2 pA/pixel)	8.64
Column (CDS) buffer amplifiers	417
Digital row circuitry	16.4
Digital column readout circuitry	20.0

5.4.13 Analysis of the centroid-tracking system

The analysis is again divided into two parts, that for the pixel and that for the edge circuitry. For the centroid-tracking pixel, sensitivity to light change and to noise will be analyzed. In addition, the linearity of the output circuit computation and the general system characteristics will be examined.

Figure 5-20 shows a schematic of the centroid-tracking pixel. *Figure 5-21* shows one cell of the edge circuitry.

5.4.14 Centroid pixel sensitivity

From section 5.4.4, the equation for the input voltage change necessary to trigger the centroid-calculating pixel is

$$\Delta V_{in} = -\frac{\Delta V_{out} U_t}{\kappa V_{0P}} \quad (5.37)$$

or

$$\Delta V_{in} = -\frac{(V_{TN} - V_{invreset}) U_t}{\kappa V_{0P}} \quad (5.38)$$

In addition to the ΔV_{in} necessary to raise the output of the inverter from its reset state to V_{TN}, the coupling of the reset switch and dummy compensation switch must also be considered. This will add a ΔV_{switch} voltage to the input that will significantly affect ΔV_{in}. While the reset switch is being turned off, it still has some finite resistance. In addition, the output of the inverter remains a low-impedance restoring voltage, which can sink enough current to offset the charge from the gate of the switch as it turns off. Therefore, most of the effect of the clock feed-through will occur as the

© M. Clapp and R. Etienne-Cummings, "Dual pixel array for imaging, motion detection and centroid tracking," *IEEE Sensors Journal*, vol. 2, no. 6, pp. 529–548, Dec. 2002 (partial reprint).

NMOS switch gate voltage goes below V_{TN}. In this region of weak inversion, most of the charge has already been depleted from the channel. The remaining clock feed-through effect resides in the gate–drain overlap capacitance. This capacitance for a minimum-size transistor is about 0.5 fF. Including bulk effects, the gate voltage at which the switch will turn off will be around 1.6 V. The charge injected by a voltage swing of 1.6 V into 0.5 fF is only about 0.8 fC. This amount of charge can be removed by the tiny subthreshold currents running through the switch before it is completely off. These currents can easily be on the order of 1 nA, even at very low gate voltages. A current of 1 nA would remove 0.8 fC in 0.8 μs. As long as the fall time of the clock signal is even slightly slow, the effect of clock feed-through will be reduced by these subthreshold currents. The charge that does feed through will arrive at the main 150 fF capacitor of the pixel. In addition, the dummy switch transistor operating in full inversion will take in more charge with its complementary clock than the main switch manages to release. It will couple into the input node with both drain and source overlap capacitances and with the gate–channel capacitance while above threshold. The combined effect of both transistors, conservatively assuming that the main switch does not conduct very well while it is in subthreshold, is as follows:

$$\Delta V_{\text{switch}} = -V_{TN}\left(\frac{C_{gd1}}{C_{\text{accap}}}\right) + \left[V_{dd}\left(\frac{C_{gd2}+C_{gs2}}{C_{\text{accap}}}\right) + \left(V_{dd}-V_{TN}\right)\left(\frac{C_{gc2}}{C_{\text{accap}}}\right)\right]$$

(5.39)

It should be noted that the "high" gate voltage of the dummy switch need not be V_{dd} as it is in this design. If the value of the logic high voltage sent to the gate of the dummy transistor were changed, the expression for ΔV_{switch} would be different. The value of this variable voltage would be substituted wherever V_{dd} appears in the current formula. In this way, one could directly control the value of ΔV_{switch} with this control voltage. This would in turn control the sensitivity of the pixel.

The completed expression for ΔV_{switch} allows us to write the full description for the change in photodiode voltage necessary to trip the inverter:

$$\Delta V_{\text{pdiode}} = \Delta V_{\text{in}} + \Delta V_{\text{switch}}$$

(5.40)

© M. Clapp and R. Etienne-Cummings, "Dual pixel array for imaging, motion detection and centroid tracking," *IEEE Sensors Journal*, vol. 2, no. 6, pp. 529–548, Dec. 2002 (partial reprint).

Normally, the biases and voltages are set such that the gain $A_{inv} = -1260$ and $\Delta V_{out} = 250$ mV. The value of ΔV_{in} thus becomes 198 µV and ΔV_{switch} is computed to be 56.5 mV. The voltage for ΔV_{pdiode} in this case is therefore dominated by ΔV_{switch}.

The photodiode voltage is regulated by V_{gs} of transistor M1. This V_{gs} is directly dependent on the photocurrent of the photodiode. If the light falling on this pixel induces a subthreshold pixel current (as it does for almost all lighting conditions), then the source voltage of M1 will change as the natural logarithm of current change. The current will have to increase by a specific multiplicative factor from its value during inverter reset to change V_{gs} by a sufficient amount. To change the source voltage by a specific ΔV_{pdiode}, the current will need to reach I_{trip} as described in the following equation:

$$I_{trip} = M_{light} I_{reset} \tag{5.41}$$

where

$$M_{light} = \exp\left(\frac{\Delta V_{pdiode}}{U_t}\right) \tag{5.42}$$

From the values for ΔV_{pdiode} given above, $M_{light} = 9.6$.

5.4.15 Centroid pixel noise

Since the output of each centroid-tracking pixel is a digital voltage, noise in the output voltage is not a concern. However, noise can degrade the operation of the pixel by introducing a random voltage component into ΔV_{in}, and hence in the factor of the increase in light level necessary to activate the pixel. In this section, the noise-induced uncertainty will be calculated.

The first noise source arises from the process of resetting the inverter. During reset, the output node of the inverter will exhibit kT/C noise due to the thermal noise in the inverter transistors and in the main pixel capacitor. The main pixel capacitor has a capacitance of 150 fF, so this reset noise will be $\langle v^2_{reset}\rangle = 26.5 \times 10^{-9}$ V^2 for an amplitude of 162 µV$_{rms}$. When the reset is released, the input will start with this level of noise.

The photodiode/NMOS bias subcircuit is easy to analyze for noise, because it is the same as an APS photodiode in perpetual reset. As such, it also exhibits kT/C noise. During normal (non-reset) operation, however, the main explicit capacitor in the pixel is floating and will not contribute to this noise figure. The photodiode parasitic capacitance on this node totals 82 fF. This gives $\langle v^2_{pdioide} \rangle = 50.5 \times 10^{-9}$ V^2 and a noise amplitude of 225 μV$_{rms}$.

These two noise sources together give a noise power to the input of the inverter of

$$\langle v^2_{reset} \rangle + \langle v^2_{pdiode} \rangle = 77 \times 10^{-9} \text{ V}^2 \qquad (5.43)$$

or an RMS noise voltage of 277.5 μV. This much voltage change on the input of the inverter corresponds to an extra factor of 1.01 of photocurrent and hence light intensity. Compared to the threshold for tripping the inverter of $M_{light} = 9.58$, this is a small amount of noise—clearly not enough to cause accidental activation of the inverter.

5.4.16   Centroid subsystem speed

The limiting speed of centroid operation is dependent on the reset time and the propagation delay of the pixel inverter. There is a certain minimum time that the inverters of the pixels need during reset to settle to their final trip point.

Each inverter during reset can be modeled as a PMOS current source, a diode-connected NMOS transistor, and a main pixel capacitor to AC ground. The diode-connected NMOS can be approximated as a resistor with a value of $1/g_{m5}$, with g_{m5} calculated at the operating point where V_{out} has reached its final equilibrium voltage. This will not be a true representation of the circuit operation, since an accurate analysis would require a large signal model. However, the approximation of $1/g_{m5}$ at the equilibrium point will give results that are always more conservative than the true behavior. If V_{out} starts high and must fall to reach equilibrium, the actual g_{m5} will be larger than the equilibrium g_{m5} and the circuit will reach equilibrium faster. If V_{out} starts lower than equilibrium, the actual g_{m5} will be lower, and again the circuit will reach equilibrium faster in reality than the approximation would indicate.

Figure 5-22(a) shows the simplified inverter circuit; *Figure 5-22(b)* is the equivalent small-signal circuit. In the small-signal model, the inverter becomes a simple RC circuit with $\tau = C_{accap}/g_{m5}$, where $g_{m5} = I_D(\kappa_N/U_t)$ for the circuit in a subthreshold state. At equilibrium, $I_D = 68.82$ pA, and therefore $g_{m5} = 2.27 \times 10^{-9}$ mho and $R = 439.8$ MΩ. With $C_{accap} = 150$ fF, this gives a time constant of $\tau = 66.0$ μs. However, $(2.2)\tau$, or 145 μs, would

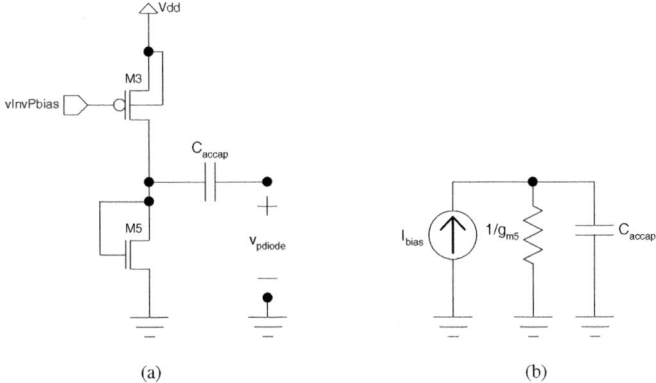

Figure 5-22. (a) Simplified centroid pixel inverter in reset. *(b)* Schematic diagram of an equivalent small-signal inverter in reset.

be a more conservative time constant, with the voltage moving 90% of its total swing in that time. The actual reset time constant will be shorter, but 145 μs will be a good conservative minimum.

The propagation delay of the inverter is directly dependent on the bias of M3. The parasitic capacitance on the output of the inverter is roughly 12.3 fF. This yields a propagation delay of

$$t_{inverter} = \frac{CV_{TN}}{I_{bias}} = \frac{(12.3\,\text{fF})(0.75\,\text{V})}{I_{bias}} \tag{5.44}$$

For $C = 68.82$ pA, $t_{inverter} = 134$ μs. Summing this with the minimum reset time gives a total minimum cycle time of 279 μs and a maximum centroid rate of 3580 Hz. It should be noted that increasing the inverter bias current by a factor of 10 will decrease the inverter propagation time by a factor of 10, but will increase the centroid system current consumption by only about 1.5%.

The time after reset and before the tripping of the inverter is spent waiting for the photocurrent to change by a sufficient magnitude. The length of this period of time, t_{detect}, determines the minimum rate of change of light levels in order to be detected:

$$R_{change} = \frac{M_{light}}{t_{detect}} \rightarrow t_{detect} = \frac{M_{light}}{R_{change}} \tag{5.45}$$

Table 5-5. Estimated current consumption of centroid-calculating circuits.

Circuit	Current Consumption
Photocurrent (6 pA/pixel)	12.96 nA
Digital circuitry (180 Hz)	305 pA
Pixel inverters (68.8 pA/pixel)	148.7 nA
Resistive ladder diffamps (1.2 μA/A)	115.2 μA
Resistive ladder	4 μA

For a desired maximum R_{change}, equation (5.46) computes the minimum t_{detect} part of the centroid cycle. The final cycle time takes

$$t_{cycle} = t_{reset} + t_{detect} + t_{inverter} \qquad (5.46)$$

Conversely, the longest one can wait without the inverter falsely tripping determines the maximum period possible with this centroid circuit. Leakage from the switch drain and source diffusions limits the amount of time the input of the inverter can remain floating. The maximum time before the leakage current alone causes the inverter to trip can be predicted given that (i) $\Delta V_{in} = 198$ μV (from section 5.4.14), (ii) the leakage current from the three drain/source diffusions is 60.75 aA, and (iii) the capacitance of the input to the inverter is 150 fF. In the absence of light, this should happen at $\Delta t_{leak} = 489$ ms.

5.4.17   Centroid power consumption

The power consumption of the centroid-tracking circuitry depends on the photocurrent drawn by the continuously biased photodiode, the operation of the pixel inverter, and the digital and analog circuitry on the periphery of the chip.

Photocurrent can easily vary by decades depending on the intensity of the incident light. A photocurrent of 6 pA may be assumed here, since it has actually been observed with this chip under indoor lighting conditions. Given this level of incident light, the continuously biased pixels use about 13 nA over the whole array. The total current of this block has been estimated to be 116 μA, of which the largest share goes to the buffer diffamps on the resistive ladder of the edge circuitry. *Table 5-5* shows the computed current draw of each portion of the centroid circuitry.

© M. Clapp and R. Etienne-Cummings, "Dual pixel array for imaging, motion detection and centroid tracking," *IEEE Sensors Journal*, vol. 2, no. 6, pp. 529–548, Dec. 2002 (partial reprint).

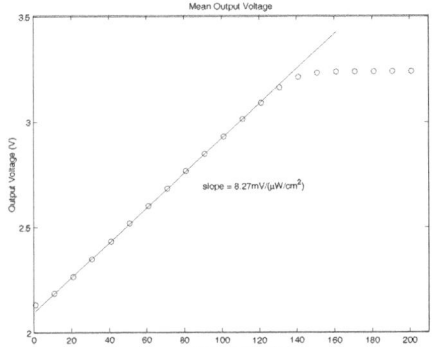
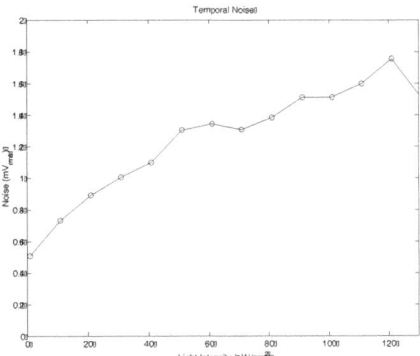

Figure 5-23. Output pixel voltage as a function of input light level for a 33 ms integration time.

Figure 5-24. Temporal noise at the output of the chip.

5.4.18 Measured APS linearity, gain, and sensitivity

A graph of pixel voltage output as a function of light intensity input is shown in *Figure 5-23*. The points corresponding to the input light intensities between 11 $\mu W/cm^2$ and 121 $\mu W/cm^2$ were fitted to the straight line shown in the figure. Within this range, the slope was 8.27 mV/($\mu W/cm^2$). For an integration time of 33 ms, this corresponds to 250 $\mu V/((\mu W/cm^2)\cdot ms)$, which differs by only 3.3% from the estimate of 244 $\mu V/((\mu W/cm^2)\cdot ms)$ in section 5.4.7.

The voltage range corresponding to this fit was 2.19 V to 3.09 V, for a linear range of 0.9 V. Within this range, the RMS error of the voltage values from the fitted line was 2.3 mV.

5.4.19 Measured APS noise

For the APS imager, all measured noise exceeded the predictions from the analysis. This is understandable considering that it was necessary to omit $1/f$ noise from the computed noise value.

Temporal noise in the APS imager was measured over time at the same pixel location, and was extremely low compared to other noise sources. All measurements indicated a temporal noise of less than 1.8 mV_{rms} over the linear signal range (see *Figure 5-24*). The APS system noise predicted in section 5.4.8 was 552 μV_{rms}, which corresponds to the minimum value in *Figure 5-24*. This measured noise was so low that it likely approached the minimum levels that our board readout electronics were capable of detecting accurately.

© M. Clapp and R. Etienne-Cummings, "Dual pixel array for imaging, motion detection and centroid tracking," *IEEE Sensors Journal*, vol. 2, no. 6, pp. 529–548, Dec. 2002 (partial reprint).

Pixel-to-pixel fixed-pattern noise for the imager was barely detectable at low light levels, but increased at an almost linear rate as the signal level increased. The large-signal equation for a gate–source follower with transistor M1 as the follower and M2 as the current source is

$$V_{in} = V_{out} + V_{T0} + \gamma_1\sqrt{2|\phi_F| + V_{out}} - \sqrt{2|\phi_F|} + \sqrt{\frac{I_{bias}}{K'_N\left(\dfrac{W}{L}\right)_1}}\sqrt{\frac{1+\lambda_2 V_{out}}{1+\lambda_1\left(V_{dd}-V_{out}\right)}}$$

$$(5.47)$$

Given equation (5.47), an expression for the change in output voltage due to a change in input voltage (as would be computed by the CDS unit) can be derived:

$$V_{in}(\text{reset}) - V_{in}(\text{final}) = V_{out}(\text{reset}) - V_{out}(\text{final}) - \gamma_1 B - \sqrt{\frac{I_{bias1}}{K'_N\left(\dfrac{W}{L}\right)_1}}\,D$$

$$B = \sqrt{2|\phi_F| + V_{out}(\text{reset})} - \sqrt{2|\phi_F| + V_{out}(\text{final})}$$

$$D = \sqrt{\frac{1+\lambda_2 V_{out}(\text{reset})}{1+\lambda_1\left(V_{dd}-V_{out}(\text{reset})\right)}} - \sqrt{\frac{1+\lambda_2 V_{out}(\text{final})}{1+\lambda_1\left(V_{dd}-V_{out}(\text{final})\right)}}$$

$$(5.48)$$

From equation (5.48), it can be seen that the deviation from a direct $\Delta V_{in} = \Delta V_{out}$ relationship involves the bulk effect of the pixel amplifier transistor (B) and the drain conductance of both pixel and column current source transistors (D). It can be shown that the factor D increases almost linearly with decreasing V_{out} given that $\lambda_1 = 0.0801$ and $\lambda_2 = 0.0626$, which is the case for this APS system. As D increases, it magnifies the contribution of the $(W/L)_1$ factor from the gate–source follower in the expression. Thus, any variation in the gate–source follower transistor geometry in the pixel will have an increasing effect as the signal increases and V_{out} decreases. To decrease this effect, λ_1 and λ_2 need to be reduced. Lengthening the column current source transistor (and also the pixel transistors if space allows) will accomplish this.

Column-to-column FPN results were relatively constant over the linear range of the imager. These stayed at or near the worst pixel-to-pixel noise of about $16\,\text{mV}_{rms}$. There are two chief reasons for this undesirable

© M. Clapp and R. Etienne-Cummings, "Dual pixel array for imaging, motion detection and centroid tracking," *IEEE Sensors Journal*, vol. 2, no. 6, pp. 529–548, Dec. 2002 (partial reprint).

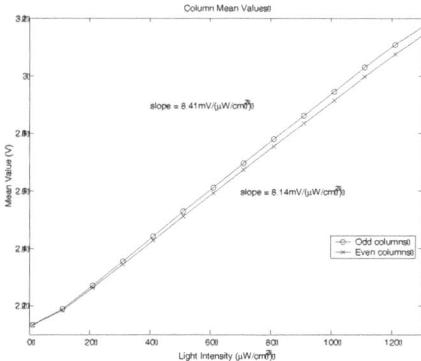

Figure 5-25. Voltage as a function of light intensity for odd and even columns.

performance. First of all, this simple architecture used a CDS circuit on the end of every column, so there was no way to correct column-to-column offsets. A global CDS circuit would alleviate much of the column-to-column FPN.

The second reason has to do with the layout of the pixel array. Because two APS pixels are tiled above and below each centroid-mapping pixel in the array, the physical and electrical environments of adjacent APS pixels are not the same. Put more simply, for every two adjacent APS pixels, one will see the left sides of the neighboring centroid pixels and the other will see the right sides of the neighboring centroid pixels. Doping profiles for left and right APS pixels will be slightly different because of the asymmetric placement of the photodiode area within the centroid pixel. This differing proximity of the photodiode will also cause the amount of photogenerated carriers in the substrate to be different for left and right APS pixels under the same incident light. These are unfortunately types of gain errors, and as such cannot be remedied by the CDS circuit. As a result, alternating lighter and darker vertical stripes are apparent in images of scenes that are dimly and evenly lit. This phenomenon is also apparent when measurements are taken separately for odd and even column groups. *Figure 5-25* clearly shows a different light–voltage transfer function for odd columns than for even columns. When the average pixel value of every column in the array is measured for an evenly lit scene, the distribution is not normal. The entire array has instead a binodal distribution, with two separate and distinct mean values especially at higher light levels. The column-to-column FPN of all odd columns and of all even columns taken separately are each better than the combined FPN figure (see *Figure 5-26*).

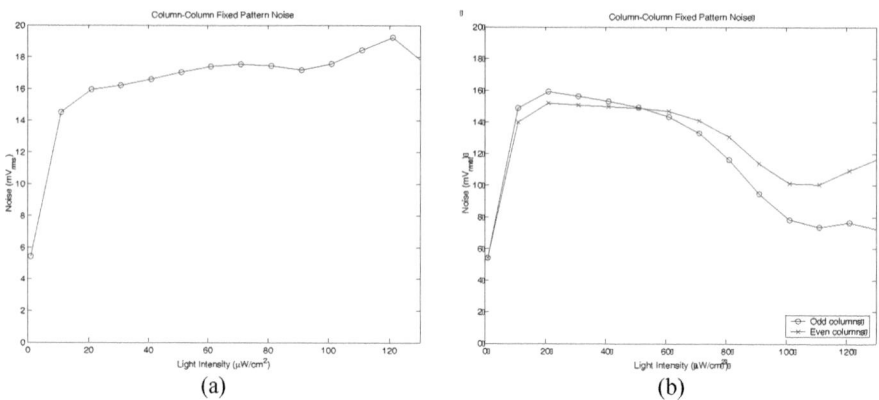

Figure 5-26. (a) Column-column fixed-pattern noise for all columns together (entire image). *(b)* Column-column fixed-pattern noise for odd and even columns taken separately.

In future chips, barrier wells could be used to isolate APS pixels from centroid pixels and improve performance, but this would be at the expense of the area and fill factor of the array. Other solutions would be to make the array perfectly regular (with one APS pixel for every centroid pixel) or to make the centroid pixels perfectly symmetric.

5.4.20   Measured APS dynamic range and SNR

The total noise of the APS imager for different light levels can be seen in *Figure 5-27*. Sample images are shown in *Figure 5-27(c)*. At maximum signal level, the total noise (standard deviation/mean signal level) of 2.88% corresponds to an SNR of 30.8 dB.

5.4.21   Measured APS speed

Imaging at a high enough frame rate to test the limits of the APS is difficult, due to the high light levels necessary for a useful signal. The readout circuitry could be tested for speed, however. These circuits still functioned properly up to a pixel clock speed of 11 MHz, or a period of 91 ns. This was also the limit of the test circuitry.

5.4.22   Centroid frequency response

In section 5.4.16, it was estimated that the fastest rate of blinking at which a stationary blinking light would remain detectable was 3.6 kHz. To

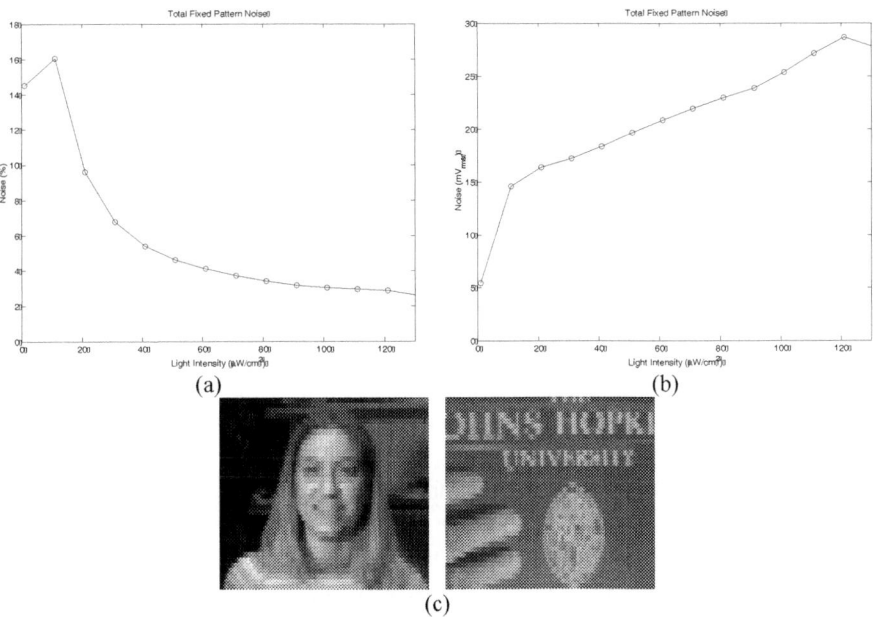

Figure 5-27. (a) Total FPN percentage (standard deviation/mean signal level) for different light levels. *(b)* Total FPN (RMS voltage) for different light levels. *(c)* Sample pictures from the APS image array.

test this calculation, a blinking LED was fed with a square-wave signal of sufficient amplitude and variable frequency. The frequency at which the blinking actually ceased to be detectable was around 4.2 kHz.

It was also calculated in the same section that the slowest possible reset time would be 489 ms. To confirm this, the array was covered (to protect it from light) and left until the pixels tripped through leakage. The actual measured time varied between 450 ms and 490 ms in the dark, and was about 200 ms in ambient light with no motion. This strongly suggests that the drain–source diffusions of the inverter switches are leaking either due to indirect light falling on the pixels or due to the effects of minority carriers in the substrate from the nearby photodiodes. Such effects could be reduced with more careful layout of the switch transistors.

© M. Clapp and R. Etienne-Cummings, "Dual pixel array for imaging, motion detection and centroid tracking," *IEEE Sensors Journal*, vol. 2, no. 6, pp. 529–548, Dec. 2002 (partial reprint).

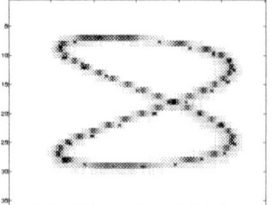
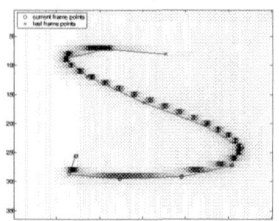
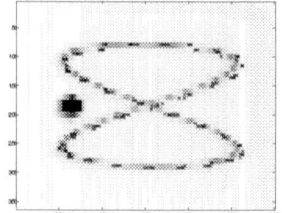

Figure 5-28. Position plot of the output data from the centroid chip superimposed on the sum of a series of reverse-video APS images.

Figure 5-29. Reverse-video image of one APS frame with 6 corresponding centroid positions plus 6 positions from the previous APS frame.

Figure 5-30. Sum of reverse-video APS images of a target moving in a figure 8 and of a stationary LED, both with all corresponding centroid positions.

5.4.23 Centroid system performance

The centroid tracking system was tested using an analog oscilloscope screen as a target. The X/Y mode setting was used, and two function generators set to 10 Hz and 20 Hz supplied the scope channels. In this way, a moving point of light tracing a stable figure-8 pattern could be observed on the oscilloscope screen. APS image data and centroid coordinate data were taken simultaneously. Centroid voltages were converted to digital data and sent to a controlling computer. A composite image of all APS frames was produced by summing all frames and then inverting the brightness of the image for easier printing. On top of this composite image was plotted the centroid positions reported by the centroid-tracking subsystem of the chip. The result is displayed in *Figure 5-28*. The data is an excellent match of the target, which was composed of two sine waves in the x and y directions. Six centroid coordinates were taken for every APS frame taken. One such APS image and the centroid coordinates of the current and previous frames are displayed in *Figure 5-29*. It is obvious that whereas the APS imager sees one smear of the path of the oscilloscope point, the centroid-tracking circuitry is able to accurately and precisely plot specific points along the path in real time.

Figure 5-30 shows another example of the cumulative centroid positions reported for an oscilloscope target. A non-blinking stationary LED was placed next to the moving oscilloscope target to demonstrate that the stationary LED had no effect on centroid positions despite it being much brighter than the oscilloscope.

© M. Clapp and R. Etienne-Cummings, "Dual pixel array for imaging, motion detection and centroid tracking," *IEEE Sensors Journal*, vol. 2, no. 6, pp. 529–548, Dec. 2002 (partial reprint).

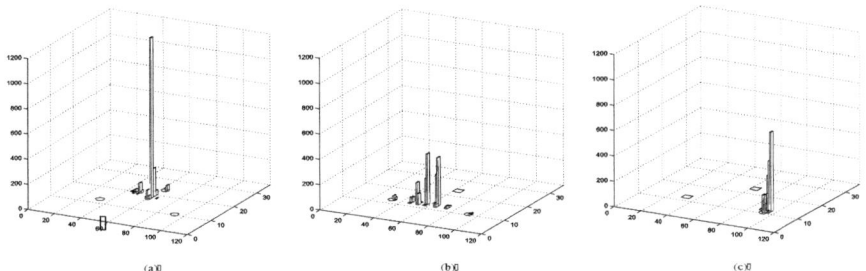

Figure 5-31. Two-dimensional histograms of centroid response with a target of three LEDs. Circles indicate blinking LED positions and squares indicate steadily on LED positions: *(a)* 3 blinking; *(b)* 2 blinking, 1 steadily on; and *(c)* 1 blinking, 2 steadily on.

With faster moving targets, the speed of the centroid subsystem could be increased even more. Centroid pixels are sensitive to changes in incident light since their last reset. Therefore, faster changes in light (faster movement) would allow for shorter reset intervals and higher measurement frequency.

In addition to trials involving a single moving target, experiments with multiple targets were performed. In the first set of experiments, a target of three LEDs in a triangle formation was imaged. All the LEDs were either blinking or steadily on, and all were stationary. Three different tests were performed. The first test had all three LEDs blinking at exactly the same time. *Figure 5-31(a)* shows a histogram of the centroid positions reported by the chip, with blinking LED positions marked by circles. From this histogram, we can see that the vast majority of positions reported are in the center of the triangle. Notice that since two LED positions are on nearly the same row, their contributions to the row position of the centroid are overlapping. Since the method for centroid determination ignores the number of active pixels in a row, the computed centroid is closer to the far point of the LED triangle than would be expected from a true centroid. The weight of the far point (row 23) in the centroid computation is comparable to both LEDs together on row 10 of the graph. The second experiment was the same as the first, except that one LED was continuously on instead of blinking. In *Figure 5-31(b)*, the non-blinking LED location is marked with a square outline instead of a circle. The positions plotted lie between the two blinking LED positions and are unaffected by the steadily on LED. Similarly, *Figure 5-31(c)* shows a test with one blinking LED position (marked with a circular outline) and two non-blinking steadily on LEDs (marked with square outlines). In this case, there is no doubt that the only

© M. Clapp and R. Etienne-Cummings, "Dual pixel array for imaging, motion detection and centroid tracking," *IEEE Sensors Journal*, vol. 2, no. 6, pp. 529–548, Dec. 2002 (partial reprint).

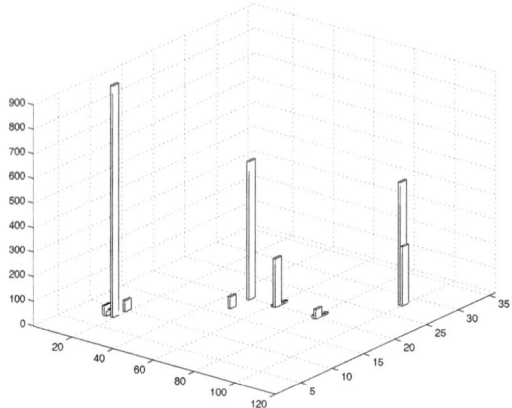

Figure 5-32. Two-dimensional histogram overlaid on the imager array, showing reported centroid positions for two LEDs with separate blinking phases and periods.

positions reported are at the only element of the scene that is changing in time.

Another experiment was performed that involved multiple LEDs with uncorrelated blinking. Two LEDs with separate blinking periods and phases were set up at different x and y positions in front of the imager, and centroid positions were recorded. *Figure 5-32* shows a histogram of the number of values recorded in specific regions of the array. In addition to the two positions of the actual LEDs showing a marked response, the linear combination of their positions also shows a considerable number of recorded coordinates. If two LEDs are seen to blink in the same period of time t_{detect}, the centroid of their positions will be computed and reported. This is the normal operation of the centroid subsystem. Multiple target tracking is still possible, however, with the addition of some basic statistical analysis of the positions reported. Through techniques such as Singular Value Decomposition (SVD), the linearly independent positions can be extracted and the linear combination of the two positions can be recognized as a false position. These techniques have more limitations in their applicability. For instance, if the true movement of an object happened to coincide with the linear combination of the movement of two other objects, it might be falsely omitted. For simple observations of a few objects, however, it is possible to extract meaningful position data for all objects involved. A system with broader applicability could be constructed by changing the edge circuitry of the centroid subsystem, allowing the detection of multiple regions of activity in the array. This hardware modification will be pursued in the next-generation imager.

© M. Clapp and R. Etienne-Cummings, "Dual pixel array for imaging, motion detection and centroid tracking," *IEEE Sensors Journal*, vol. 2, no. 6, pp. 529–548, Dec. 2002 (partial reprint).

5.5   Conclusions

This chapter has presented current-mode, voltage-mode, and mixed-mode focal-pane image processing approaches. The current-mode approach readily allows the outputs of pixels to be scaled and summed, and thus very compact and fast spatial convolutions can be realized on the focal plane. Linearity and fixed-pattern noise are, however, usually worse in current-mode imagers than in voltage-mode imagers. For temporal filtering and processing, voltage-mode imaging is better because correlated double sampling immediately provides a vital component of most temporal processors: the temporal differencer. Hence, a pipeline method was developed for implementing an image temporal differencing scheme that equalizes the delay between the two images for all pixels. This approach also allowed fine control on the magnitude of the delay, and can be used to increase the frame access rate. Both of these systems used a computation-on-readout (COR) architecture, which involves a block-parallel, sequential processing of the pixels. Lastly, a mixed-mode imager was described that used both voltage-mode imaging and processing (for low-noise, high-pixel-density imaging) and current-mode imaging (for motion detection and centroid localization). This mixed-mode imager also combined pixel-serial and pixel-parallel processing. The impact of this combination is an increase in FPN, because the environment seen by some pixels is different from that seen by others. The phototransduction gain is also negatively affected in the mixed-mode imager. Nonetheless, all three modes can be effectively used for low-power and small-footprint focal-plane image processing architectures.

ACKNOWLEDGMENTS

This work has been supported by NSF CAREER Award #9896362, NSF award number #ECS-001–0026, NSF ERC on Computer Integrated Surgical Systems and Technology at Johns Hopkins University, ONR YIP Award #N000140010562, and a collaboration grant between Johns Hopkins University and the Army Research Laboratory at Adelphi, MD.

BIBLIOGRAPHY

[1]   O. Yadid-Pecht, R. Ginosar, and Y. S. Diamand, "A random access photodiode array for intelligent image capture," *IEEE Trans. Electron Devices*, vol. 38, pp. 1772–1780, 1991.

[2]   S. K. Mendis, S. E. Kemeny, and E. R. Fossum, "A 128 × 128 CMOS active pixel image sensor for highly integrated imaging systems," in *Proc. IEEE Int. Electron Devices Meeting*, 1993, pp. 22.6.1–22.6.4.

[3]   B. Ackland and A. Dickinson, "Camera on a chip," in *Proc. IEEE Int. Solid-State Circuit Conf.*, 1996, pp. 22–25.

[4]   E. R. Fossum, "CMOS image sensor: electronic camera-on-a-chip," *IEEE Trans. Electron Devices*, vol. 44, pp. 1689–1698, 1997.

[5]   C. Mead and M. Ismail, Eds. *Analog VLSI Implementation of Neural Networks*. Newell, MA: Kluwer Academic Press, 1989.

[6]   W. Camp and J. Van der Spiegel, "A silicon VLSI optical sensor for pattern recognition," *Sensors and Actuators A*, vol. 3, no. 1–3, pp. 188–195, 1994.

[7]   K. Boahen, "Retinomorphic chips that see quadruple images," in *Proc. Microneuro '99*, Granada, Spain, 1999.

[8]   K. Boahen and A. Andreou, "A contrast sensitive silicon retina with reciprocal synapses," in *Advances in Neural Information Processing Systems*, Vol. 4, J. Moody, J. Hanson, and R. Lippmann, Eds. San Mateo, CA: Morgan Kaufmann, 1992, pp. 764–772.

[9]   C. Koch and H. Li, Eds. *Vision Chips: Implementing Vision Algorithms with Analog VLSI Circuits*. IEEE Computer Society Press, 1995, pp. 139–161.

[10] B. E. Shi, "A low-power orientation-selective vision sensor," *IEEE Trans. on Circuits and Systems II: Analog and Digital Signal Processing*, vol. 47, no. 5, pp. 435–440, 2000.

[11] K. B. Cho, B. Sheu, and W. Young, "Biologically inspired image sensor/processor architecture with 2-D cellular neural network for vision," in *Proc. 1998 IEEE Int. Joint Conf. Neural Networks*, 1998, Part 1 (of 3), pp. 569–574.

[12] R. Dominguez-Castro, S. Espejo, A. Rodriguez-Vazquez, R. Carmona, P. Foldesy, A. Zarandy, P. Szolgay, T. Sziranyi, and T. Roska, "A 0.8 μm CMOS two-dimensional programmable mixed-signal focal-plane array processor with on-chip binary imaging and instructions storage," *IEEE J. Solid-State Circuits*, vol. 32, no. 7, pp. 1013–1026, 1997.

[13] A. Zarandy, M. Csapodi, and T. Roska, "20 μsec focal plane image processing," in *Proc. 6th IEEE Int. Workshop on Cellular Neural Networks and their Applications*, 2000, pp. 267–271.

[14] B. Feher, P. Szolgay, T. Roska, A. Radvanyi, T. Sziranyi, M. Csapodi, K. Laszlo, L. Nemes, I. Szatmari, G. Toth, and P. Venetianer, "ACE: a digital floating point CNN emulator engine," in *Proc. IEEE Workshop on Cellular Neural Networks and their Applications*, 1996, pp. 273–278.

[15] C. B. Umminger and C. G. Sodini, "Switched capacitor networks for focal plane image processing systems," *IEEE Trans. Circuits and Systems for Video Technologies*, vol. 2, no. 4, pp. 392–400, 1992.

[16] S. Mendis et al., "CMOS active pixel image sensors for highly integrated imaging systems," *IEEE J. Solid-State Circuits*, vol. 32, no. 2, pp. 187–197, 1997.

[17] C. Sodini, J. Gealow, Z. Talib, and I. Masaki, "Integrated memory/logic architecture for image processing," in *Proc. 11th Int. Conf. VLSI Design*, 1998, pp. 304–309.

[18] M. Schwarz, R. Hauschild, B. Hosticka, J. Huppertz, T. Kneip, S. Kolnsberg, L. Ewe, and Hoc Khiem Trieu, "Single-chip CMOS image sensors for a retina implant system," *IEEE Trans. Circuits and Systems II*, vol. 46, no. 7, pp. 870–877, 1999.

[19] H. Yamashita and C. Sodini, "A 128×128 CMOS imager with 4×128 bit-serial-parallel PE array," in *Proc. IEEE Int. Solid-State Circuit Conference*, 2001, pp. 96–97.

[20] R. Geiger, P. Allen, and N. Strader, *VLSI: Design Techniques for Analog and Digital Circuits*. New York, NY: McGraw-Hill, 1990.

[21] A. Andreou, K. Boahen, P. Pouliquen, and A. Pavasovic, "Current mode sub-threshold MOS circuits for analog VLSI neural systems," *IEEE Trans. Neural Networks*, vol. 2, no. 2, pp. 205–213, 1991.

[22] Y. Degerli, F. Lavernhe, P. Magnan, and J. A. Farre, "Analysis and reduction of signal readout circuitry temporal noise in CMOS image sensors for low-light levels," *IEEE Trans. Electron Devices*, vol. 47, no. 5, pp. 949–962, 2000.

[23] B. Horn, *Robot Vision*. Cambridge, MA: MIT Press, 1986.

[24] R. Etienne-Cummings, V. Gruev, and M. Abdel Ghani, "VLSI implementation of motion centroid localization for autonomous navigation," *Adv. Neural Information Processing Systems,* vol. 11, pp. 685–691, 1999.

[25] R. Etienne-Cummings and Donghui Cai, "A general purpose image processing chip: orientation detection," *Adv. Neural Information Processing Systems,* vol. 10, pp. 873–879, 1998.

[26] R. Gonzalez and R. Woods, *Digital Image Processing*. Reading, MA: Addison-Wesley Publishing Company, 1992.

[27] G. Strang and T. Q. Nguyen, *Wavelets and Filter Banks,* rev. ed. Wellesley, MA: Wellesley-Cambridge Press, 1998.

[28] M. A. Vorontsov, G. W. Carhart, M. H. Cohen, and G. Cauwenberghs, "Adaptive optics based on analog parallel stochastic optimization: analysis and experimental demonstration," *J. Opt. Soc. Am. A*, vol. 17, no. 8, pp. 1440–1453, 2000.

[29] A. Dickinson, B. Ackland, E.-S. Eid, D. Inglis, and E. R. Fossum, "A CMOS 256×256 active pixel sensor with motion detection," in *Proc. IEEE Int. Solid State Circuits Conference*, 1996, pp. 226–227.

[30] S. Ma and L. Chen, "A single-chip CMOS APS camera with direct frame difference output," *IEEE J. Solid-State Circuits,* vol. 34, pp. 1415–1418, Oct. 1999.

[31] S. Mendis, S. E. Kemeny, E. R. Fossum, "CMOS active pixel image sensor," *IEEE Trans. Electron Devices*, vol. 41, no. 3, pp. 452–453, Mar. 1994.

[32] M. H. White, D. R. Lampe, F. C. Blaha, and I. A. Mack, "Characterization of surface channel CCD image arrays at low light levels," *IEEE J. Solid-State Circuits*, vol. SC-9, no. 1, pp. 1–13, Feb. 1974.

[33] C. Enz and G. Temes, "Circuit techniques for reducing the effects of op-amp imperfections: autozeroing, correlated double sampling, and chopper stabilization," *Proc. IEEE*, vol. 84, no. 11, pp. 1584–1614, Nov. 1996.

[34] K. Boahen, "Retinomorphic chips that see quadruple images," in *Proc. Microneuro '99*, Granada, Spain, Mar. 1999.

[35] K. A. Boahen, "Point-to-point connectivity between neuromorphic chips using address events," *IEEE Trans. Circuits and Systems II: Analog and Digital Signal Processing,* vol. 47, no. 5, pp. 416–434, May 2000.

[36] E. Culurciello, R. Etienne-Cummings and K. Boahen, "Arbitrated address event representation digital image sensor," in *IEEE Int. Solid-State Circuits Conf. 2001. [Dig. Tech. Papers,* pp. 92 –93].

[37] H. Barlow, *The Senses: Physiology of the Retina*. Cambridge, MA: Cambridge Univ. Press, 1982.

[38] R. H. Nixon, S. E. Kemeny, B. Pain, C. O. Staller, and E. R. Fossum, "256 × 256 CMOS active pixel sensor camera-on-a-chip," *IEEE J. Solid-State Circuits*, vol. 31, no. 12, pp. 2046–2050, 1996.

[39] P. E. Allen and D. R. Holberg, *CMOS Analog Circuit Design*. Oxford: Oxford University Press, 1987, pp. 197–321, 365–455.

[40] C. D. Motchenbacher and J. A. Connelly, *Low-Noise Electronic System Design*. New York: John Wiley & Sons, 1993, pp. 1–78, 140–171.

Chapter 6

CMOS IMAGER NON-UNIFORMITY CORRECTION USING FLOATING-GATE ADAPTATION

Marc Cohen[1] and Gert Cauwenberghs[2]
[1]Institute for Systems Research, University of Maryland
College Park, MD 20742, USA
[2]Department of Electrical and Computer Engineering, Johns Hopkins University
Baltimore, MD 21218, USA

Abstract: Stochastic adaptive algorithms are investigated for on-line correction of spatial non-uniformity in random-access addressable imaging systems. The adaptive architecture is implemented in analog VLSI, and is integrated with the photo sensors on the focal plane. Random sequences of address locations selected with controlled statistics are used to adaptively equalize the intensity distribution at variable spatial scales. Through a logarithmic transformation of system variables, adaptive gain correction is achieved through offset correction in the logarithmic domain. This idea is particularly attractive for compact implementation using translinear floating-gate MOS circuits. Furthermore, the same architecture and random addressing provide for oversampled binary encoding of the image resulting in an equalized intensity histogram. The techniques apply to a variety of solid-state imagers, such as artificial retinas, active pixel sensors and IR sensor arrays. Experimental results confirm gain correction and histogram equalization in a 64 × 64 pixel adaptive array integrated on a 2.2 mm × 2.25 mm chip in 1.5 µm CMOS technology.

Key words: On-line correction, non-uniformity correction, adaptation, equalization, floating-gate, focal plane, CMOS imager, analog VLSI.

6.1   Introduction

Since the seminal work by Carver Mead on neuromorphic floating-gate adaptation in the silicon retina [1], few groups have addressed the problem of on-line adaptive correction of non-uniformities on the focal plane in solid-state image sensor arrays [2, 3] and neuromorphic vision sensors [4, 5]. Most efforts have concentrated instead on non-adaptive correction using on-chip [6] or off-chip calibrated storage. Gain and offset non-uniformities in the photo sensors and active elements on the focal plane contribute "salt-and-pepper" fixed-pattern noise at the received image, which limits the resolution and sensitivity of imaging systems and image postprocessing. Flicker noise and other physical sources of fluctuation and mismatch make it necessary to correct for all these non-uniformities *on-line,* which is problematic since the image received is itself unknown. Existing "blind" adaptive algorithms for on-line correction are complex and the amount of computation required to implement them is generally excessive. Integration on the focal plane would incur a significant increase in active pixel size, a decrease in spatial resolution, a decrease in fill factor of the imager, and an increase in power consumption.

In this work, a class of *stochastic* adaptive algorithms has been developed that integrate general non-uniformity correction with minimal, if not zero, overhead in the number of active components on the focal plane. In particular, floating-gate adaptive CMOS technology is used to implement a two-transistor adaptive-gain element for on-line focal-plane compensation of current gain mismatch. The algorithms make effective use of the statistics of pixel intensity under randomly selected sequences of address locations, and avoid the need for extra circuitry to *explicitly* compute spatial averages and locally difference the result. The resulting stochastic algorithms are particularly simple to implement.

The stochastic algorithms for adaptive non-uniformity correction that take advantage of the spatial statistics of image intensity can also be used to perform image intensity equalization and normalization on the focal plane. Equalization is a useful property because it maximizes the available dynamic range and assigns higher sensitivity to more statistically frequent intensities. At the same time, the image is converted into digital form, thus avoiding the need for *explicit* analog-to-digital conversion.

In this chapter, the stochastic algorithms for adaptive non-uniformity correction are formulated. A simple logarithmic transform on offset correction allows the use of the same algorithms for gain correction. Intensity equalization is discussed as a natural extension of these stochastic rules. The floating-gate translinear current-mode VLSI implementation of the adaptive pixel is described and analyzed. The system architecture is also described, including the external circuits used for experimental validation of the VLSI imager. Experimental results for gain non-uniformity correction and image intensity equalization are discussed.

6.2   Adaptive non-uniformity correction

Non-uniformity correction can be approached using two strategies: apply a uniform reference image to the static imager and ensure that all pixel outputs are equal [1], or drift natural scenes across the imager where each pixel subtracts its output from its spatially low-pass filtered output to derive an error signal [7]. The former is referred to as *static* non-uniformity correction (SNUC) and the latter as *scene-based* non-uniformity correction (SBNUC). Our imager can accommodate either type of mismatch correction strategy. The SBNUC algorithm has been implemented on the focal plane in CMOS- and IR-based imagers [2], and has been successful in reducing offset mismatch.

In this chapter, SNUC will be used as the primary method to reduce current gain mismatch in a phototransistor-based CMOS imager or silicon retina. An adjustable, adaptive pixel current gain can be achieved by applying a controllable voltage offset on a floating-gate transistor in each pixel. The system architecture also allows SBNUC through control of the statistics of random address sequences.

First, the problem must be set up in terms of established on-line algorithms for offset correction. Then this same algorithm can be extended to gain mismatch reduction through a simple logarithmic transformation of system state variables.

Figure 6-1(a) schematically demonstrates the offset correction technique. The set of system equations is

$$y = x + o, \quad z = y + q = x + o + q, \tag{6.1}$$

where x is the input random (sensor) variable, y is the received input with unknown offset o, q is the applied offset correction, and z is the corrected output. For offset cancellation,

$$o + q \equiv constant \; \forall \; \text{pixels}. \tag{6.2}$$

A simple (gradient descent) adaptive rule [7] to achieve this is

$$\Delta q = -\alpha \left(z - z_{ref} \right), \tag{6.3}$$

which adjusts the output z on average towards a reference z_{ref}.

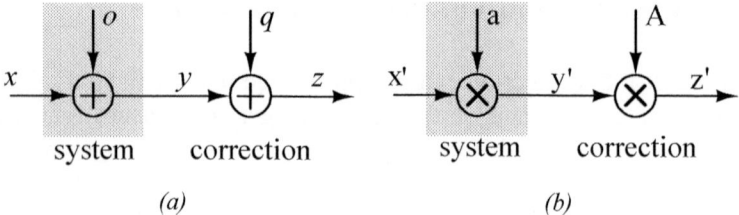

Figure 6-1. (a) Offset correction. (b) Gain correction. © 2001 IEEE.

The reference is constructed by expressing

$$z_{ref} = \begin{cases} \langle z \rangle & \text{for SNUC} \\ \langle z \rangle_{local} & \text{for SBNUC} \end{cases} \tag{6.4}$$

where the $\langle \, \rangle$ symbol represents spatial averaging at global and local scales, respectively, and α denotes the adaptation (or "learning") rate. Circuits implementing a locally differenced diffusive kernel (with adjustable space constant) to perform the computations in equation (6.3) are presented in [2]. One can introduce a stochastic version of this rule

$$\Delta q_{\mathbf{r}(k)} = -\alpha \left(z_{\mathbf{r}(k)} - z_{\mathbf{r}(k-1)} \right) \tag{6.5}$$

where the subscripts $\mathbf{r}(k-1)$ and $\mathbf{r}(k)$ denote pixel addresses at consecutive time steps $(k-1)$ and k respectively. Taking expectations on both sides of equation (6.5) (for a particular pixel selected at time k) yields

$$E\left[\Delta q_{\mathbf{r}(k)} \right] = -\alpha \left(z_{\mathbf{r}(k)} - E\left[z_{\mathbf{r}(k-1)} \right] \right)$$

which depends on the statistics of the consecutive address selections as determined by the conditional transition probabilities (densities) given by $p(\mathbf{r}(k-1) \,|\, \mathbf{r}(k))$. Therefore, by controlling the statistics through proper choice of the random sequence of addresses $\mathbf{r}(k)$ [i.e., $p(\mathbf{r}(k-1) \,|\, \mathbf{r}(k))$], one can implement, on average, the spatial convolution kernels needed for both SNUC and SBNUC in equation (6.4). In particular, for a random sequence with the terms $\mathbf{r}(k-1)$ and $\mathbf{r}(k)$ independent [i.e., $p(\mathbf{r}(k-1) \,|\, \mathbf{r}(k)) = p(\mathbf{r}(k-1))$],

$$\mathrm{E}\left[z_{\mathbf{r}(k-1)} \right] = \langle z \rangle \qquad (6.6)$$

whereas if $\mathbf{r}(k-1)$ and $\mathbf{r}(k)$ are related by embedding memory in the address sequence (for example, through inertia or by imposing limits on $\Delta\mathbf{r} = \mathbf{r}(k) - \mathbf{r}(k-1)$),

$$\mathrm{E}\left[z_{\mathbf{r}(k-1)} \right] = \langle z \rangle_{local}. \qquad (6.7)$$

Equation (6.5) is a stochastic on-line version of SNUC, and likewise equation (6.7) implements stochastic SBNUC. Hardware requirements can be further simplified by thresholding the update in equation (6.5) to a signed version of that rule (a "pilot" rule),

$$\Delta q_{\mathbf{r}(k)} = -\alpha \operatorname{sign}\left(z_{\mathbf{r}(k)} - z_{\mathbf{r}(k-1)} \right) \qquad (6.8)$$

with fixed-size update increments and decrements.

6.3   Canceling gain non-uniformity

The gradient descent formulation [7] also adaptively compensates for gain mismatch, although it does not prevent the gain from becoming negative. The approach in this chapter is to relate gain correction (under the positivity constraint imposed by current-domain circuits) to offset correction through a logarithmic transformation. This transformation has a physical meaning that can be exploited in the hardware implementation as discussed in the next section. *Figure 6-1(b)* schematically illustrates the concept of gain mismatch correction in relation to *Figure 6-1(a)*.

The system is governed by the equations

$$y' = a\,x', \quad z' = A\,y' = A\,a\,x', \qquad (6.9)$$

which can be transformed into

$$\ln z' = \ln A + \ln a + \ln x', \qquad (6.10)$$

such that for gain non-uniformity correction,

$$\ln A + \ln a \equiv constant \ \forall \ \text{pixels}. \qquad (6.11)$$

By identifying corresponding terms (in particular, $\ln A = q$ (or $A = e^q$) and $\ln a = o$) in equations (6.1) and (6.10), and because of the monotonicity of the logarithmic map, the learning rule of equation (6.8) can be rewritten as

$$\Delta q_{\mathbf{r}(k)} = -\alpha \operatorname{sign}\left(z'_{\mathbf{r}(k)} - z'_{\mathbf{r}(k-1)} \right), \tag{6.12}$$

which in turn can be expressed as a stochastic on-line learning rule with *relative* gain increments:

$$\Delta A_{\mathbf{r}(k)} = -\alpha A_{\mathbf{r}(k)} \operatorname{sign}\left(z'_{\mathbf{r}(k)} - z'_{\mathbf{r}(k-1)} \right). \tag{6.13}$$

6.4   Intensity equalization

The corrections in the *constant* terms both in the offset equation (6.2) and in the gain equation (6.11) are undefined and not regulated during the adaptation. This problem can be circumvented by properly normalizing the acquired image. One particularly attractive approach to normalization is to equalize the image intensity histogram, which in addition to mapping the intensity range to unity also produces a maximum entropy coded output [8]. Incidentally, the same stochastic algorithms of equations (6.8) and (6.13) for non-uniformity correction can also be used for histogram-equalized image coding. Pixel intensities are mean-rate encoded in a single-bit over sampled representation akin to delta-sigma modulation [9], although without the need for integration or any other processing at the pixel level. This could be compared with a popular scheme for neuromorphic multi-chip systems, the address-event communication protocol [10], in which sparse pixel-based events such as spiking action potentials are communicated asynchronously across chips. In the technique described here, addresses are not event-based, but are supplied synchronously with prescribed random spatial statistics.

In particular, the image is coded in terms of the bits obtained by comparing $z_{\mathbf{r}(k)}$ and $z_{\mathbf{r}(k-1)}$ as in equation (6.8) or $z_{\mathbf{r}'(k)}$ and $z_{\mathbf{r}'(k-1)}$ as in equation (6.13). If larger, a '1' symbol is transmitted; otherwise, it is a '0'. The selected address is either part of the transmitted code or it is generated at the receiver end from the same random seed. Thus, the code is defined as

$$f\left(z_{\mathbf{r}(k)} \right) = \operatorname{sign}\left(z_{\mathbf{r}(k)} - z_{\mathbf{r}(k-1)} \right). \tag{6.14}$$

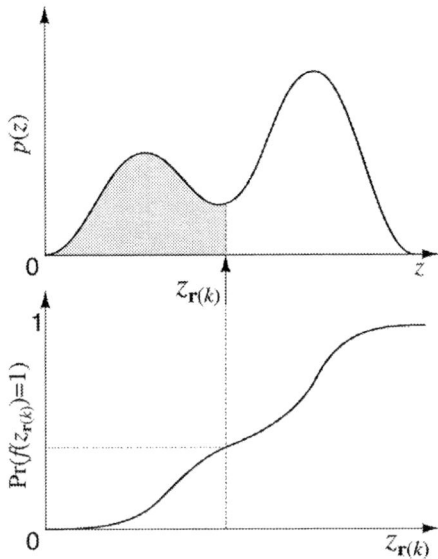

Figure 6-2. Input intensity probability density function (top), and corresponding mean-rate transfer function (bottom) for intensity equalization and normalization. © 2001 IEEE.

By selecting random addresses with controlled spatial statistics as in equation (6.4), this code effectively compares the intensity of a selected pixel with a base value that is either a global or a local average. The probability of '1' is the fraction of pixels in that neighborhood with intensity lower than the present pixel. This is illustrated in *Figure 6-2*, in which the mean-rate pixel activity is given by the cumulative probability density function

$$\Pr\left(f\left(z_{\mathbf{r}(k)}\right)=1\right)= \int_{0}^{z_{\mathbf{r}(k)}} p(z)dz. \tag{6.15}$$

This corresponds to intensity equalization and normalization of the image, a desirable feature for maintaining a large dynamic range in image acquisition [11]. As seen in *Figure 6-2*, the coding transfer function assigns higher sensitivity to statistically more frequent intensities. The uniform distribution and *maximum entropy* encoding obtained by this transformation is a well-known result and appears to take place in biological phototransduction as well [8]. The mechanism of image equalization as achieved here is unique in that it evolves from statistical techniques in an oversampled representation, and the statistics of the address sequence can be tailored to control the size of the neighborhood for different spatial scales of intensity normalization.

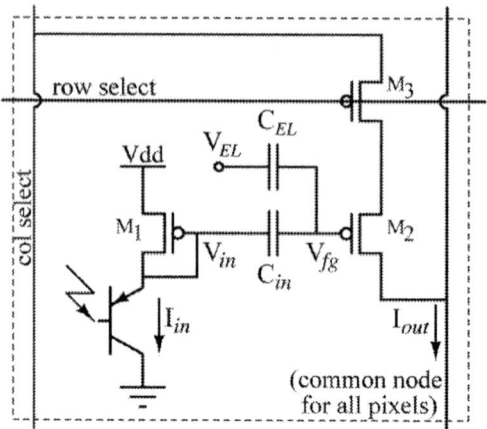

Figure 6-3. Circuit diagram of the floating-gate adaptive pixel. © 2001 IEEE.

6.5   Focal plane VLSI implementation

Rather than implementing equation (6.13) directly, the exponential relationship between voltage and current in a (subthreshold) MOS transistor is used to encode a current *gain* as the exponential of a differential voltage across a *floating-gate* capacitor. The increments and decrements $\pm\Delta q$ in equation (6.13) are then naturally implemented by hot electron injection and tunneling across the floating-gate oxide [12]. The voltage on the floating gate is then a function of the charge,

$$V_{fg} = \lambda \left(V_{EL} + \frac{Q}{C_{EL}} \right) + (1-\lambda)V_{in} \qquad (6.16)$$

where Q is the charge injected or tunneled onto or from the floating gate, $\lambda = C_{EL}/(C_{EL} + C_{in}) \approx 0.3$, and V_{EL} is an externally applied global voltage for all pixels. The schematic of the floating-gate pixel is shown in *Figure 6-3*. A vertical pnp bipolar transistor converts photon energy to emitter current I_{in} with current gain β. Transistors M_1 and M_2 form a floating-gate current mirror with adjustable gain [13]. The output current I_{out} of the pixel is sourced by transistor M_2 and measured off-chip. The gate and source of transistor M_3 provide random access pixel addressing at the periphery as needed to implement the stochastic kernel. For this pixel design, equation (6.16) establishes the following current transfer function in the subthreshold regime:

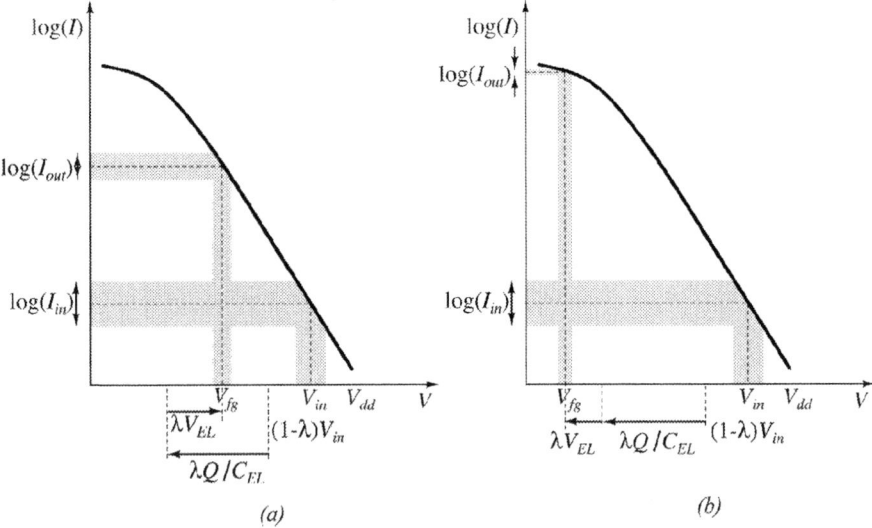

Figure 6-4. Pictorial interpretation of the contributions of λ, V_{EL} and Q/C_{EL} to the pixel current transfer function *(a)* for subthreshold output current and *(b)* for above-threshold output current. © 2001 IEEE.

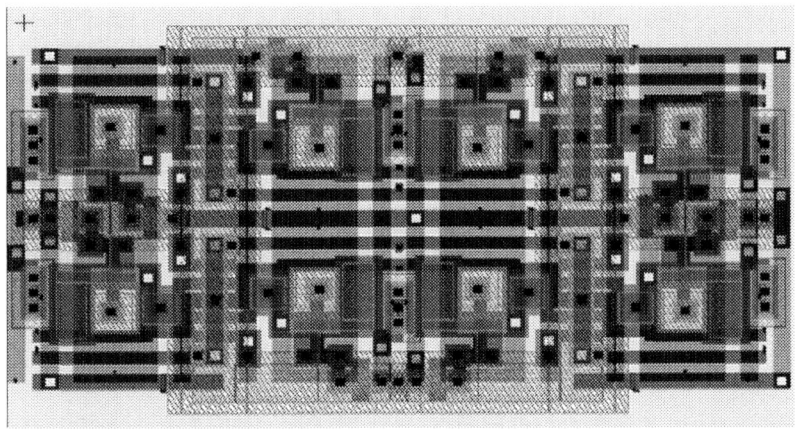

Figure 6-5. Layout of a 2 × 4 pixel sub-array.

$$I_{out} = c \, (I_{in})^{1-\lambda} \exp\left(\frac{-\kappa\lambda Q}{C_{EL}V_T}\right) \exp\left(\frac{-\kappa\lambda V_{EL}}{V_T}\right) \tag{6.17}$$

where $c = (I_0 W/L \exp(V_{dd}/V_T))^\lambda$, I_0 is the subthreshold leakage current, W and L are the width and length of transistors M_1 and M_2, V_{dd} is the supply voltage, V_T is the thermal voltage, and κ is the subthreshold slope factor.

The first exponential factor on the right in equation (6.17) corresponds to the adaptive gain correction A, while the second exponential factor represents normalization, which is globally controlled by V_{EL}. By injecting electrons onto (or tunneling electrons from) the floating gate [12], Q is incrementally (or decrementally) altered, which in turn logarithmically modulates A and thereby effectively implements the pilot rule of equation (6.12).

Figure 6-4 illustrates the effect of the various contributions to the pixel current transfer function I_{out} through the floating-gate voltage V_{fg} as given by equation (6.17). Capacitive division between C_{in} and C_{EL} reduces the voltage swing on the floating gate V_{fg} by a factor $(1 - \lambda)$ relative to the input voltage V_{in}. Through the logarithmic *V*-to-*I* transformation across the MOS transistor for subthreshold output current, this factor compresses the dynamic range of intensities in the output image,

$$I_{out} = c'(I_{in})^\varepsilon , \tag{6.18}$$

by a factor $\varepsilon = (1 - \lambda)$, as shown in *Figure 6-4(a)*. Hot electron injection onto the floating gate modulates the charge Q, and thereby corrects the (relative) gain in each pixel individually by correspondingly lowering the floating-gate voltage V_{fg}. The electrode voltage V_{EL} allows for a global shift of V_{fg} for all pixels, in either a positive or negative direction as shown in *Figure 6-4(a)* and *6-4(b)*. The effect either way is a global, electronically adjustable scale factor in the gain, which allows for automatic gain control. For lower values of V_{EL}, which bring the transistor M_2 above the threshold as indicated in *Figure 6-4(b)*, a smaller compression factor ε is obtained in the current transfer function although this factor ε then depends on the signal. If the image is subsequently histogram equalized through the oversampled binary encoding, the nonlinearity in the transfer function ε becomes irrelevant.

The layout of a 2×4 pixel sub-array is shown in *Figure 6-5*. Note that while the layout minimizes pixel size, it does so at the expense of analog transistor matching. The consequences of this can be seen in the "before correction" image in *Figure 6-8*.

6.6　　VLSI system architecture

An array of 64×64 adaptive pixels was integrated with x and y random-access addressing decoders onto a 2.2 mm $\times$ 2.25 mm chip in 1.5 µm CMOS technology. A photomicrograph of the prototype fabricated through MOSIS is shown in *Figure 6-6*.

Figure 6-7 illustrates the architecture of the chip and the setup used to experimentally validate the concept of reducing the gain mismatch between pixels on the prototype adaptive array.

The imager was uniformly illuminated and a single column and row address $\mathbf{r}(x(k), y(k)) = \mathbf{r}(k)$ was randomly selected. With switch S_1 closed and S_2 open, $I_{out}(k)$ was measured using a transimpedance amplifier to generate a voltage $z_{\mathbf{r}(k)}$. If $f(z_{\mathbf{r}(k)}) = 0$, S_1 was opened and S_2 momentarily closed. The drain of transistor M_2 was pulsed down to $V_{inj} \approx (V_{dd} - 8V)$ and a small packet of negative charge was injected onto the floating gate. If $f(z_{\mathbf{r}(k)}) = 1$, the gain of the selected pixel was not altered and the process was continued by randomly selecting a new pixel.

A *one-sided* version of the stochastic learning rule of equation (6.13) was implemented:

$$\Delta q_{(k)} = \begin{cases} \alpha & f\left(z_{(k)}\right) < 0; \\ 0 & \text{otherwise.} \end{cases} \tag{6.19}$$

Because adaptation is active in only one direction, the average level $\langle z \rangle$ drifts in that direction over time. The coupling electrode voltage V_{EL} can be used to compensate for this drift and implement automatic gain control.

After gain non-uniformity correction, the imager can be used to acquire static natural images. Using random addresses with prescribed statistics, the output bit from the comparator $f(z_{\mathbf{r}(k)})$ can also be accumulated in bins whose addresses are defined by $\mathbf{r}(k)$. The resulting histogram then represents the intensity-equalized acquired image.

6.7　　Experimental results

The 64×64 phototransistor-based imager was uniformly illuminated using a white light source. The pixel array was scanned before any gain mismatch correction and again after every 200 cycles of correction, until the correction was judged to be completed after 2800 cycles. Each of the 4096 pixels was selected in random sequence every cycle. *Figure 6-8* shows the

Figure 6-6. Photomicrograph of the 64 × 64 pixel adaptive imager chip. Dimensions are 2.2 mm × 2.25 mm in 1.5 µm CMOS technology. © 2001 IEEE.

evolution of the histograms built from the I_{out} recorded from each pixel on the focal plane versus adaptation cycle number. Also shown are the scanned images from the chip before and after gain mismatch correction.

The standard deviation of I_{out} (i.e., $\sigma_{I_{out}}$) normalized to the mean $\langle I_{out} \rangle$ was measured and plotted versus $\langle I_{out} \rangle$ before and after gain mismatch correction. *Figure 6-9* plots these experimental results. The five different $\langle I_{out} \rangle$ values correspond to five different levels of illumination, which have been labeled 1, 2, 3, 4, and 5. Adaptation was done at the illumination level corresponding to label 5.

A black and white 35-mm slide was projected onto the imager after gain mismatch correction and the array was scanned. The slide contained a light grey character "R" against a dark grey background (both bitmapped). The resulting image, as scanned from the imager chip, is shown in *Figure 6-10*.

A 35-mm grayscale (pixilated 64 × 64) image of an eye, shown in *Figure 6-11(a)*, was also projected onto the imager. The acquired image is shown in *Figure 6-11(b)*, and the histogram-equalized image (which was obtained from a 256-times-oversampled binary coding of the chip output) is shown in *Figure 6-11(c)*.

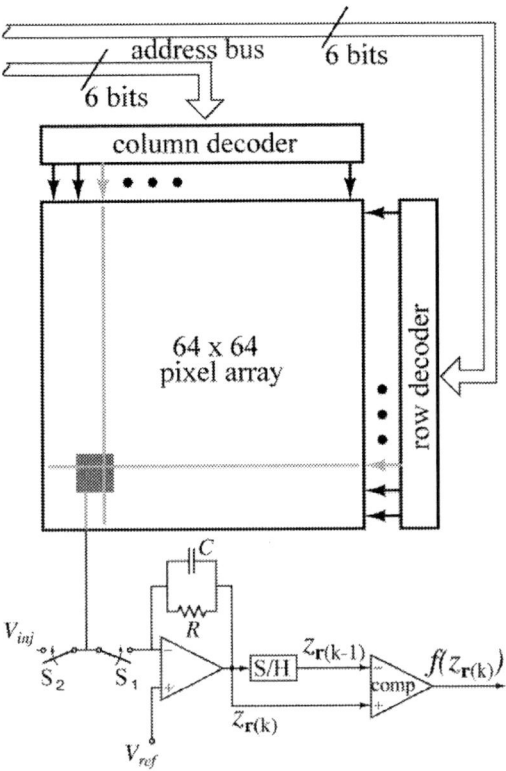

Figure 6-7. Chip architecture and system setup for gain mismatch correction and intensity histogram equalization. © 2001 IEEE.

6.8   Discussion

Injecting a negative packet of charge onto the floating gate of transistor M_2 lowers its gate voltage and therefore increases its output current. Consequently, correction is in one direction only, increasing the current gain. Since the efficiency of charge injection depends exponentially on the magnitude of drain-source current through the device [12], pixels having higher I_{out} will inject more each time their drains are pulled down to V_{inj}. This positive feedback mechanism can be kept in check either by driving the source of the floating-gate p-FET transistor with a current source (as shown in *Figure 6-12*), or by setting V_{inj} appropriately, keeping S_2 closed for a fixed time interval ($\approx 20\ \mu s$), and having hysteresis in the comparator that computes $f(z_{\mathbf{r}(k)})$. The latter option was chosen here to provide simplicity in the test setup.

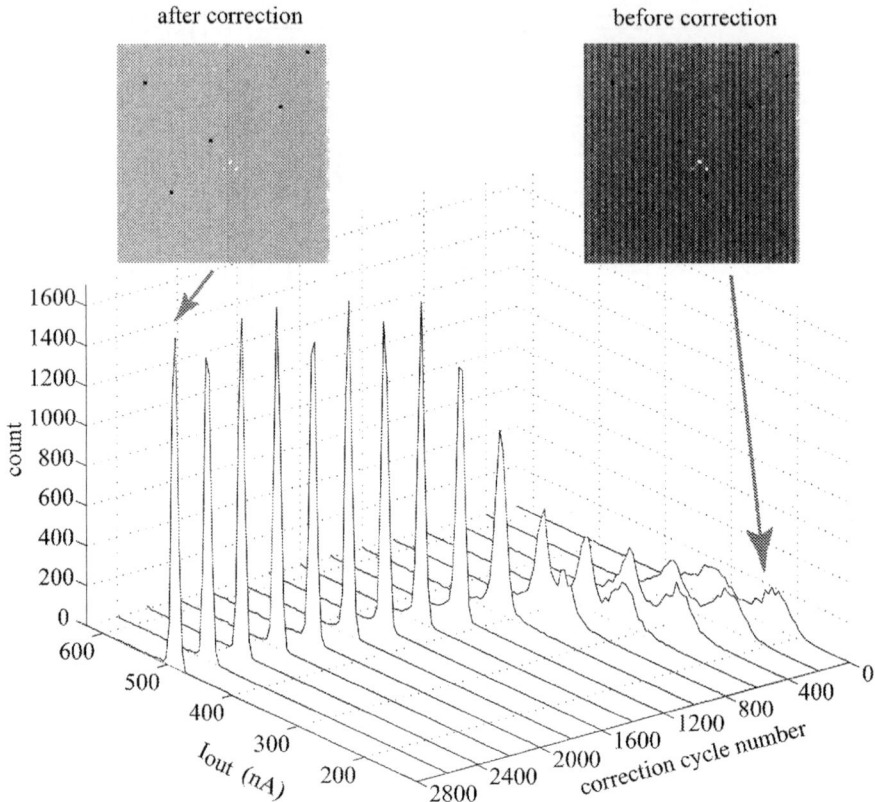

Figure 6-8. The time course of gain non-uniformity reduction as recorded from the adaptive imager chip. Also shown are images acquired before and after gain correction was performed under conditions of uniform illumination. © 2001 IEEE.

Figure 6-12 shows three different schemes for performing hot electron injection onto the floating gate of a p-FET transistor. The first depicts the method used in the chip presented here, which as explained above can lead to unstable behavior unless the necessary precautions are taken. The second method uses a current source to set the current (I_{set}) that the p-FET transistor must source. The compliance of the current source must be such that it can sink I_{set} at the appropriate drain voltage (V_{inj}) of the floating-gate p-FET transistor. The current I_{out} will approach I_{set} asymptotically and V_{inj} will rise appropriately, so as to decrease injection. The third method is the one chosen for implementation in the latest version of this chip. Each time the drain of the floating-gate p-FET transistor is pulsed down to V_{inj}, its output current I_{out} will increase (almost) linearly, and the rate at which it increases can be set by I_{inj} and the pulse width. I_{inj} is common for all pixels in the array.

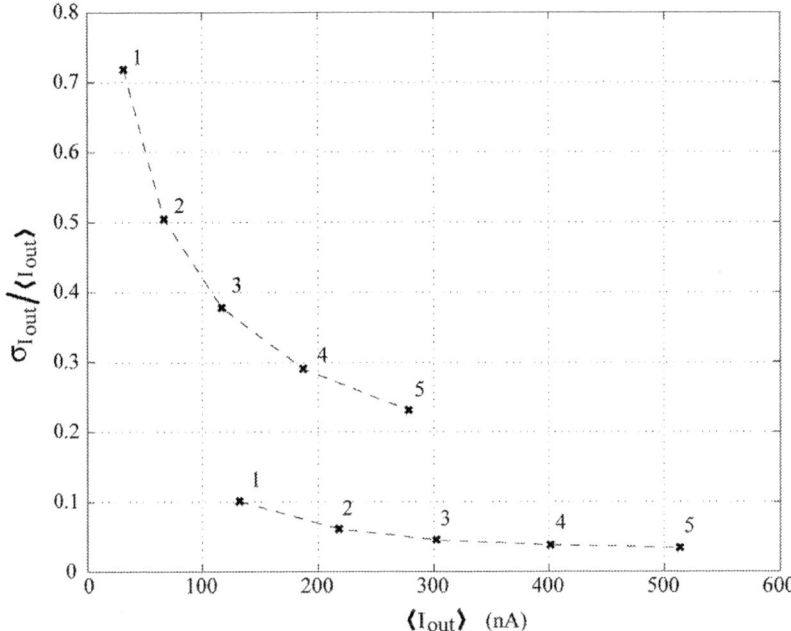

Figure 6-9. Experimental $\sigma_{I_{out}}/\langle I_{out}\rangle$ versus $\langle I_{out}\rangle$ for five different illumination intensities before gain correction (top curve) and after gain correction (bottom curve). © 2001 IEEE.

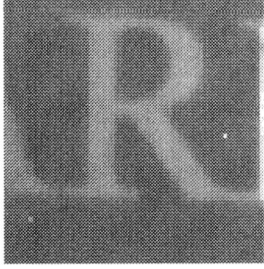

Figure 6-10. Example image acquired from the adaptive imager chip after gain mismatch reduction. A light-grey letter "R" against a dark-grey background was projected onto the chip. © 2001 IEEE.

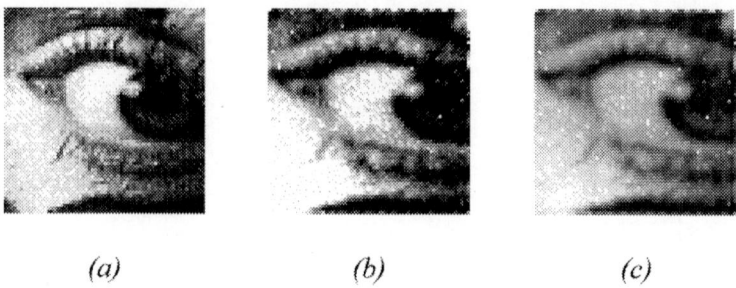

(a)   *(b)*   *(c)*

Figure 6-11. (a) Original image, *(b)* image acquired from the chip, and *(c)* equalized image obtained from oversampled binary coding (binning) of the chip outputs. © 2001 IEEE.

The "before correction" scanned image in *Figure 6-8* shows strong vertical striations in I_{out}. After the gain mismatch correction procedure, these striations are no longer visible. However, five dark pixels (low I_{out}) can be seen in this image. These pixels are "stuck" off and therefore experience negligible injection when they are selected. In the new version of the floating-gate imager, the current source I_{inj} prevents any pixels from staying in this "stuck" state. Ideally, an impulse would be expected in the histogram after correction, with all pixels having the same I_{out} when uniformly illuminated. In reality, a single narrow peak is seen in the histogram due to the injection efficiency being proportional to current and due to hysteresis in the comparator.

Figure 6-9 demonstrates that gain mismatch (and not just $\sigma_{I_{out}}/\langle I_{out}\rangle$) was reduced as a consequence of increasing $\langle I_{out}\rangle$ [14]. The pre- and post-correction data lie on two separate curves, demonstrating that there is indeed a dramatic reduction in gain mismatch due to adaptation. At low $\langle I_{out}\rangle$ (i.e., low illumination), there is a reduction in $\sigma_{I_{out}}/\langle I_{out}\rangle$ from 70% to 10%. At higher $\langle I_{out}\rangle$ (i.e., high illumination), the reduction is from 24% to 4%.

The scanned image of an "R" after adaptation shown in *Figure 6-10* gives a clear image mostly free of gradients and other fixed pattern noise present in the imager before compensation. The remaining "salt and pepper" noise (two pixels in *Figure 6-10*) is an artifact of the inhomogeneous adaptation rates under voltage-controlled hot electron injection in the setup of *Figure 6-6*, which can be alleviated by using the current-controlled setup of *Figure 6-12*. The "eye" image after intensity equalization in *Figure 6-11(c)* reveals more (intensity) detail, especially around the iris, than the acquired image in *Figure 6-11(b)*.

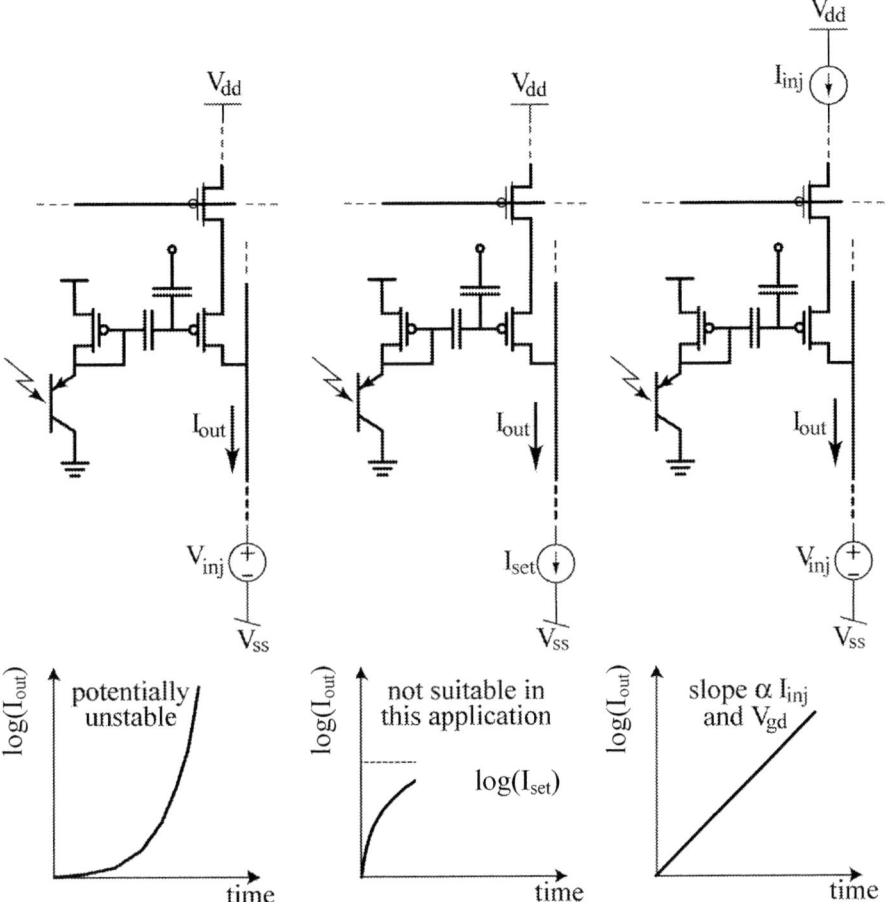

Figure 6-12. Three schemes for injecting charge onto the floating gate of the p-FET transistor in this pixel.

6.9 Conclusions

A compact pixel design and a strategy for reducing the gain mismatch inherent in arrays of phototransistors used in CMOS imagers have been introduced. It has been shown how the learning rule for offset correction can be transformed into the logarithmic domain to produce a stable learning rule for on-line gain mismatch correction. This rule is very naturally implemented by a simple translinear circuit. The pixel incorporates a floating-gate transistor that can be incrementally injected with a small packet of negative charge. The injected charge increases the current gain of the pixel in relative terms (i.e., by constant increments on a logarithmic scale).

Experimental results from a custom 64×64 phototransistor-based adaptive-pixel CMOS array (fabricated through MOSIS) prove that this pixel design and learning rule were successful for SNUC. In addition, intensity histogram equalization and digital coding of the output image were demonstrated in a binary oversampled representation, by means of the same random-addressing stochastic algorithms and architecture that was used for the adaptation.

ACKNOWLEDGMENTS

We greatly appreciate IEEE's permission to use text and figures from reference [3]. We thank Chris Diorio, Bradley Minch, Paul Hassler and Kwabena Boahen for discussions at lectures and panels. The chip was fabricated through the MOSIS foundry service.

BIBLIOGRAPHY

[1] C. A. Mead, "Adaptive retina," in *Analog VLSI Implementations of Neural Systems*, Mead and Ismail, Eds. Norwell, MA: Kluwer Academic, 1989.

[2] P. O. Pouliquen, A. G. Andreou, C. W. Terrill and G. Cauwenberghs, "Learning to compensate for sensor variability at the focal plane," in *Proc. Int. Conf. Neural Networks (IJCNN)*, Washington, DC, July 1999.

[3] M. Cohen and G. Cauwenberghs, "Floating-gate adaptation for focal-plane online nonuniformity correction," *IEEE Trans. Circuits and Systems II*, vol. 48, no. 1, pp. 83–89, 2001.

[4] Y. Chiang and J. G. Harris, "An analog integrated circuit for continuous-time gain and offset calibration of sensor arrays," *Analog Integrated Circuits and Signal Processing*, vol. 1, pp. 231–238, 1997.

[5] A. Pesavento, T. Horiuchi, C. Diorio, and C. Koch, "Adaptation of current signals with floating-gate circuits," in *Proc. 17th Int. Conf. on Microelectronics for Neural, Fuzzy and Bio-inspired Systems*, pp. 128–134, 1999.

[6] A. Aslam-Siddiqi, W. Brockherde, M. Schanz, and B. J. Hosticka, "A 128-pixel CMOS image sensor with integrated analog nonvolatile memory," *IEEE J. Solid-State Circuits*, vol. 10, pp. 1497–1501, 1998.

[7] D. A. Scribner, K. A. Sarkady, M. R. Kruer, J. T. Caulfield, J. D. Hunt, M. Colbert, and M. Descour, "Adaptive retina-like preprocessing for imaging detector arrays," in *Proc. Int. Conf. Neural Networks (ICNN)*, Boulder, CO, Feb. 1993, pp. 1955–1960.

[8] S. B. Laughlin and D. Osorio, "Mechanisms for neural signal enhancement in the blowfly compound eye," *J. Exp. Biol.*, vol. 144, pp. 113–146, 1989.

[9] J. Nakamura, B. Pain, T. Nomoto, T. Nakamura, and E. R. Fossum, "On-focal-plane signal processing for current-mode active pixel sensors," *IEEE Trans. Electron Devices,* vol. 10, pp. 1747–1758, 1997.

[10] K. A. Boahen, "A retinomorphic vision system," *IEEE Micro,* vol. 5, pp. 30–39, 1996.

[11] Y. Ni, F. Devos, M. Boujrad, and J. H. Guan, "Histogram-equalization-based adaptive image sensor for real-time vision," *IEEE J. Solid-State Circuits,* vol. 7, pp. 1027–1036, 1997.

[12] P. Hasler, B.A. Minch, J. Dugger, and C. Diorio, "Adaptive circuits and synapses using pFET floating-gate devices," in *Learning on Silicon,* G. Cauwenberghs and M. Bayoumi, Eds. Norwell, MA: Kluwer Academic, 1999.

[13] H. Miwa, K. Yang, P. Pouliquen, N. Kumar, and A. Andreou, "Storage enhancement techniques for digital memory based, analog computational engines," in *IEEE Int. Symp. on Circuits and Systems,* vol. 5, pp. 45–49, 1994.

[14] A. Pavasovic, A. G. Andreou, and C. R. Westgate, "Characterization of subthreshold MOS mismatch in transistors for VLSI systems," *J. VLSI Signal Processing*, Kluwer Academic, vol. 8, pp. 75–85, 1994.

Appendix: List of Symbols

$1/f$	1/frequency
1P3M	CMOS process with 1 poly and 3 metal layers
α	absorption coefficient [cm^{-1}]
	learning rate
A	photodiode active area [μm^2]
A, A_{inv}	DC gain
a	lattice constant
	unknown gain in multiplicative system
A_D	photodiode area, bottom component [μm^2]
A_{eff}	effective photoactive area of a pixel [μm^2]
$A_{\mathbf{r}(k)}$	gain at pixel address $\mathbf{r}$ at time k
$\Delta A_{\mathbf{r}(k)}$	gain increment at pixel address $\mathbf{r}$ at time k
B	bandwidth
c	velocity of light [m/s^2]
	constant multiplier for pixel current
C_{accap}	AC coupling capacitor
C_{colum}	column capacitance
CDS	correlated double sampling
C_{EL}	capacitance of the floating gate control electrode
Cg	conversion gain [V/C]
C_{gb}	gate-bulk MOSFET capacitance

C_{gd}	gate-drain MOSFET capacitance
C_{gs}	gate-source MOSFET capacitance
C_{gc}	gate-channel MOSFET capacitance
C_{in}	capacitance of the floating gate input
C_{int}	integration capacitance [F]
C_{J0B}	zero-bias capacitance of the bottom depletion component [F/m^2]
C_{J0sw}	zero-bias capacitance of the side-wall depletion component [F/m]
C_{jdep}	depletion capacitance of the p-n junction [F]
C_{ox}	capacitance of gate oxide per unit area
C_{pdiode}	photodiode capacitance [F]
$C_{storage}$	storage capacitance [F]
CTF	contrast transfer function
$\Delta\mathbf{r}$	change in pixel address from time $(k-1)$ to time k
ΔV	change in voltage
D	diffusion coefficient [cm^2/s]
d	junction depth [μm]
D^*	specific detectivity
DDS	difference double sampling
D_n	diffusion coefficient of electrons [cm^2/s]
D_p	diffusion coefficient of holes [cm^2/s]
DR	dynamic range [dB]
ϵ	value added to the average of all input values to produce the threshold value [V]
ε	dynamic range compression factor for subthreshold pixel current
ε_{Si}	permittivity of silicon
E[x]	expected value of x
E_C	minimum conduction band energy
E_F	Fermi energy
E_g	bandgap energy
E_i	intrinsic energy level
E_V	maximum valence band energy

EXP	scaling factor
Φ_0	flux of photons
$\Phi(\lambda)$	spectral power density [W/cm^2]
φ_B	build-in potential of the bottom depletion component [V]
φ_{Bsw}	build-in potential of the sidewall depletion component [V]
ϕ_F	Fermi potential
$f(z_{\mathbf{r}(k)})$	nonlinear decision function acting on the output of a pixel
Δf	bandwidth
FF	fill factor [%]
FPN	fixed pattern noise [V]
γ_N, γ_P	threshold voltage body-effect coefficient
g_{ds}	transistor drain-source transconductance
g_m	transistor channel transconductance
g_{mb}	transistor back-gate transconductance
GOPS	giga operations per second
η	quantum efficiency
$\eta(\lambda)$	quantum efficiency for photoelectron generation [electrons/photon]
η_{CC}	charge collection coefficient
η_{CT}	charge transfer coefficient
h	Planck constant [J·s]
$\hbar$	reduced Planck constant
$h(x)$	impulse response
h_{fe}	dynamic common emitter current gain
h_{FE}	static common emitter current gain
$\hbar\omega$	photon energy
$I(i,j)$	individual pixel current in an array
$I(X;Y)$	mutual information of random variables X and Y
I_0	process-dependent and gate-geometry-dependent factors relating fundamental currents of devices
	subthreshold leakage current

I_B	current due to background light
I_C	collector current
I_{CEO}	collector-emitter current with open base
$\langle i_{\mathrm{col}}^2 \rangle$	total noise current in the column line
I_D	dark current
	transistor drain current
I_{dark}	dark current of a photodiode
I_{del}	delayed image pixel current
I_E	emitter current
$\langle i_{\mathrm{f}}^2 \rangle$	$1/f$ noise current
I_{in}	pixel photocurrent
$I_{leakage}$	reverse diode leakage current
I_{org}	original image pixel current
I_{out}	pixel output current
$\langle I_{out} \rangle$	mean of pixel output currents
I_{ph}	primary photocurrent
$I_{\mathrm{PH}}, i_{photo}$	photocurrent [A]
I_{reset}	reset current
I_{set}	set constant current sourced by the p-FET during injection
$\langle i_{\mathrm{th}}^2 \rangle$	thermal noise current
I_{trip}	photocurrent required to trigger a change in the output state of a comparator
I_x	pixel currents in summed in the x-direction
I_y	pixel currents in summed in the y-direction
J_{diff}	diffusion current density
J_{dr}	drift current density
κ	subthreshold slope factor
κ_N, κ_P	subthreshold slope factors of MOSFETs
k	discrete time step
k, K	Boltzmann constant [J/K]
$\vec{k}$	wavevector
$k_{\{p,n\}}$	process parameters of MOSFETs $(\mu_o C_{ox})$

k_1	number of electrons collected by a unit of photodiode area per unit time [μm^{-2}]
k_2	number of electrons collected by a unit of sidewall collecting surface within the substrate per unit time [μm^{-2}]
k_3	zero-bias capacitance of the bottom depletion component [$aF \cdot \mu m^{-2}$]
k_4	zero-bias capacitance of the sidewall depletion component [$aF \cdot \mu m^{-1}$]
kfps	kilo frames per second
kT, KT	thermal energy
λ	wavelength [μm]
	multiplicative factor determined by a capacitor ratio
λ_c	cutoff wavelength
L	channel length of transistors
	MOSFET channel length
L_{diff}	characteristic diffusion length [μm]
L_{eff}	characteristic effective diffusion length
L_n	electron diffusion length [μm]
L_p	hole diffusion length [μm]
μ	mobility
μ_n	electron mobility
μ_p	hole mobility
M	modulation factor
m	particle mass
m^*	effective mass
Man	mantissa [V]
MCF	modulation contrast function
MCI	modulation contrast image
MCO	modulation contrast object
m_j	grading coefficient of the bottom component of the junction depletion
m_{jsw}	grading coefficient of the sidewall component of the junction depletion

M_n	MOSFET transistor number n
$\langle v^2_{amp} \rangle$	amplifier noise voltage
$\langle v^2_{apstotal} \rangle$	total APS noise voltage
$\langle v^2_{cdsclamp} \rangle$	correlated-double-sampling clamp source noise voltage
$\langle v^2_{colcds} \rangle$	column correlated-double-sampling noise voltage
$\langle v^2_{dark} \rangle$	pixel dark noise voltage
$\langle v^2_{pdiode} \rangle$	photodiode noise voltage
$\langle v^2_{photon} \rangle$	photon noise voltage
$\langle v^2_{pixreset} \rangle$	pixel reset noise voltage
$\langle v^2_{reset} \rangle$	reset noise voltage
N	read noise [V]
n	electron concentration
	minority carriers concentration [cm^{-3}]
	row number
N_A	acceptor doping concentration
N_C	effective density of states in the conduction band
NEP	noise equivalent power
$Nf(x)$	photon flux transmitted to the pixel surface at point x [photons/s·cm^2]
n_i	intrinsic electron concentration
NOP	no operations
n_p	electron concentration in p-type semiconductor [cm^{-3}]
$N_{p\lambda}$	photon irradiance, or the number of incoming photons per unit time [photons/s]
n_{p0}	thermal equilibrium electron concentration in a p-type semiconductor [cm^{-3}]
N_V	effective density of states in the valence band
Nwell, n-well	N-type doped well in a CMOS process
ω_0	modulation frequency
(ω_x, ω_y)	spatial frequencies in the x and y directions $[= 2\pi f_x, 2\pi f_y]$
o	unknown offset in additive system
OTF	optical transfer function $[= \tau(\omega_x, \omega_y)]$
ψ_s	surface potential

P	photodiode perimeter [μm]
p	hole concentration
p	pitch size [μm]
$p(\mathbf{r}(k-1) \mid \mathbf{r}(k))$	probability of pixel address $\mathbf{r}$ at time $(k\text{-}1)$ given $\mathbf{r}$ at time k
$p(x)$	probability distribution of random variable x
$p(z)$	probability distribution of z
P_D	photodiode perimeter [μm]
p_n	hole concentration in an n-type semiconductor
p_{n0}	equilibrium hole concentration in an n-type semiconductor
P_{ph}	optical power
P_{ph}'	optical power density
$\Pr(f(z_{\mathbf{r}(k)}))$	cumulative probability density function of $f(z)$
Q	integrated collected photocharge [C]
	charge injected onto or tunneled off the floating-gate
q	electron charge or unit electrical charge [C]
	offset correction applied to additive system
QE	quantum efficiency [%]
$\Delta q_{\mathbf{r}(k)}$	change in offset correction at address $\mathbf{r}$ and time k
R	resistance
$\mathbf{r}(k)$	pixel address at time k
r_{frame}	frame rate
R_{ph}	responsivity
σ_{Iout}	standard deviation of pixel output currents
σ_s^2	normalized optical signal power
S	part of the pixel area: the area of substrate surrounding the photodiode that is unoccupied by logic elements [μm^2]
	pixel saturation level [V]
$s(x', y')$	spread function
$S(\omega_x, \omega_y)$	Fourier transform of the spread function $s(x, y)$
SNR	signal-to-noise ratio [dB]
SR(λ)	spectral response [A/W]

τ	carrier lifetime [s]
	minority carrier lifetime [s]
τ_n	electron (minority carrier) lifetime in a p-type substrate [s]
τ_p	hole lifetime [s]
T	absolute temperature [K]
T'_{row}	modified row readout time [s]
t_{clk}	clock period
T_{comp}	time needed for comparison to the threshold level [s]
T_{copy}	time needed to copy one row into the readout buffer [s]
t_{detect}	detection cycle period
T_{frame}	frame time [s]
t_{int}, t_{int}, T_{int}	integration time [s]
t_{reset}	reset cycle period
T_{row}	total time for row readout [s]
T_{scan}	time needed to scan each pixel [s]
U_t, U_T	thermal voltage
V	voltage swing [V]
$V-$	control for epsilon value decrease [V]
$V(\lambda)$	pixel signal output for a particular wavelength [V]
$V+$	control for epsilon value increase [V]
Value	actual pixel value [V]
V_{average}	average of all input values [V]
V_{bias}	bias voltage
V_d	voltage applied to the photodiode [V]
V_{DD}, V_{dd}	positive power supply voltage [V]
V_{ds}, V_{ds}	drain-source voltage of a MOSFET [V]
V_{EL}	voltage applied to the floating gate control electrode
V_{fg}	floating gate voltage
VG	virtual ground
V_G, V_g	gate voltage of a MOSFET
V_{gs}, V_{gs}	gate-source voltage of a MOSFET [V]
V_{in}	voltage at top poly layer of a floating gate p-FET

V_{inj}	voltage at which hot electron injection occurs
V_{out}	output voltage
V_{ox}	voltage drop across oxide
V_{PH}, V_{pdiode}	photodiode voltage [V]
V_{ref}	reference voltage [V]
V_{rms}	root-mean-square voltage
V_s	source voltage of a MOSFET
V_{ss}	negative power supply voltage
V_T	thermal voltage
	threshold voltage of a MOSFET [V]
V_T, V_{TO}, $V_{T(sb)}$	threshold voltage of a MOSFET [V]
W	MOSFET channel width
W/L	transistor width-to-length ratio
W_w	representation of the wth point of decision in WDR algorithm
X	constant greater than 1
x	input random (sensor) variable to additive system
x'	input random (sensor) variable to multiplicative system
y	received input with unknown offset from additive system
y'	received input with unknown offset from multiplicative system
z	corrected output from additive system
$\langle z \rangle$	spatial average at a global scale
z'	corrected output from multiplicative system
$\langle z \rangle_{local}$	spatial average at a local scale
$z_{\mathbf{r}}'(k)$	output of multiplicative system at pixel address $\mathbf{r}$ at time k
z_{ref}	reference output

Index

240

Printed by Printforce, the Netherlands